Student Solutions Manual
for

David S. Moore, George P. McCabe, and Bruce A. Craig's
Introduction to the Practice of Statistics

Eighth Edition

Patricia Humphrey
Georgia Southern University

W. H. Freeman and Company
A Macmillan Higher Education Company
New York

© 2014, 2012, 2009, 2006 by W. H. Freeman and Company

ISBN-13: 978-1-4641-3361-9
ISBN-10: 1-4641-3361-1

Printed in the United States of America.

Second Printing 2015

W. H. Freeman and Company
41 Madison Avenue
New York, NY 10010
Houndmills, Basingstoke RG21 6XS, England
www.whfreeman.com

CONTENTS

About these solutions

The solutions that follow were prepared by Patricia Humphrey. In some cases, solutions were based on those prepared for earlier editions of *IPS*. Jackie Miller reviewed the solutions, especially focusing on those which were new (or revised) from previous editions. In spite of the care that went into that process, I might have missed a subtle change in an exercise that should have resulted in a change in the solution. Should you discover any errors or have any comments about these solutions (or the odd answers, in the back of the text), please report them to me:

> Patricia Humphrey
> Georgia Southern University
> Statesboro GA 30460-8093
> e-mail: phumphre@georgiasouthern.edu

The Student Solutions Guide is intended to provide complete solutions to all of the odd-numbered exercises, except for those for which no definitive answer can be given, such that those that ask students to look up and choose data on the Internet. In the case of simple exercises, this might mean that the "solution" is identical to the answer given in the back of the text. For more complicated exercises, the solution includes more detail about the *process* of determining the answer.

To create this guide, the odd-numbered solutions were extracted from the Instructor's Guide (which contains solutions to *all* exercises). **Some editing was done after this extraction, but a few comments aimed at instructors may still remain in these solutions.**

These solutions were prepared using: primarily Minitab and R, along with SAS (for Chapter 16) and some freeware and shareware software (G•Power, and GLMStat).

The solutions given to the applet exercises, and the sample output screens, were based on the current versions of the applets at the time the solutions were written. As revisions are made to these applets, the appearance of the output screens (and in some cases, the answers) may change. Screenshots were taken on a computer running Windows 7, but that should have no significant impact on the appearance.

Using the table of random digits

Grading SRSs chosen from Table B is complicated by the fact that students can find some creative ways to (mis)use the table. Some approaches are not mistakes, but may lead to different students having different "right" answers. Correct answers will vary based on

• The line in the table on which they begin (you may want to specify where you started if the text does not specify).
• Whether you start labels, e.g., 00 or 01.
• Whether you assign labels across the rows or down the columns (nearly all lists in the text are alphabetized down the columns—this is the approach I used).

Some approaches can potentially lead to wrong answers. Mistakes to watch out for include the following:

• Students may forget that all labels must be the same length (e.g., assigning labels such as 0, 1, 2, . . . , 9, 10, . . . rather than 00, 01, 02, . . .).
• In assigning multiple labels, they may not give the same number of labels to all units. For example, if there are 30 units, students may try to use up all the two-digit numbers, thus assigning four labels to the first 10 units and only three to the remaining 20.

As an alternative to using the random digits in Table B, students can pick a random sample by generating (pseudo)-random numbers, using software (like Excel) or a calculator. With many, if not all, calculators, the sequence of random numbers produced is determined by a "seed value" (which can be specified by the user). Rather than pointing students to a particular line of Table B, you could specify a seed value for generating random numbers, so that all students would obtain the same results (if all are using the same model of calculator). The problem with this approach is that there will be no "correct" answer.

On a TI-84 or nSpire, for example, after executing the command 0→rand, the rand command will produce the sequence (rounded to four decimals) 0.9436, 0.9083, 0.1467, . . ., while 1→rand initiates the sequence 0.7456, 0.8559, 0.2254, . . . So to choose, say, an SRS of size 10 from 30 subjects, use the command 0→rand to set the seed, and then type 1+30*rand, and press ENTER repeatedly. Ignoring the decimal portion of the resulting numbers, this produces the sample
29, 28, 5, 15, 13, 23, 2, 11, 30, 7
(Generally, to generate random numbers from 1 to n, use the command 1+n*rand and ignore the decimal portion of the result.)

Using statistical software
The use of computer software or a calculator is a must for all but the most cursory treatment of the material in this text. Be aware of the following considerations:

• *Standard deviations:* Students may be confused by software that gives both the so-called "sample standard deviation" (the one used in the text) and the "population standard deviation" (dividing by n rather than $n - 1$). Symbolically, the former is usually given as s and the latter as σ (sigma), but the distinction is not always clear. For example, many computer spreadsheets have a command such as "STDEV(. . .)" to compute standard deviations, but you may need to check the manual to find out which kind it is.
    As a quick check: For the numbers 1, 2, 3, $s = 1$ while σ is approximately 0.8165. In general, if two values are given, the larger one is s and the smaller is σ. If only one value is given, and it is the wrong one, use the relationship $s = \sigma \sqrt{\dfrac{n}{n-1}}$.

• *Stemplots:* The various choices one can make in creating a stemplot (e.g., rounding or truncating the data) have already been mentioned. Minitab opts for truncation over rounding, so all of the solutions in this guide show truncated-data stemplots (except for exercises that instructed students to round). This usually makes little difference in the overall appearance of the stemplot.

• *Significant digits in these solutions:* Most numerical answers in these solutions (and in the odd-numbered answers in the back of the text) are reported to one more significant digit than the data (for example, if all the data values are integers, I used 1 decimal place); Normal probabilities are typically reported to four decimal places, and z (or t) statistics to two places. Reporting additional decimal places in answers gives an observer a feeling of accuracy that is not warranted from the original data, because those decimal places are strictly a result of dividing and taking square roots. Four decimal places have been used in most probabilities due to that convention being used in most tables.

• *Quartiles and five-number summaries:* Methods of computing quartiles vary between different packages. Some use the approach given in the text (that is, $Q1$ is the median of all the numbers below the location of the overall median, etc.), while others use a more complicated approach. For the numbers 1, 2, 3, 4, for example, we would have $Q1 = 1.5$ and $Q3 = 3.5$, but Minitab reports these as 1.25 and 3.75, respectively, while Excel reports 1.75 and 3.25.

In these solutions (and the odd-numbered answers in the back of the text), I opted to report five-number summaries are they would be found using the text's method. Generally, all quartile methods are "close," depending on the sample size. Do not be overly concerned if your technology does not match exactly what is used in the text (although Crunchit! and TI calculators DO use the text's method).

• *Boxplots:* Because quartiles may not match your text's method exactly, this may cause a difference in whether a data value is flagged as a "suspected" outlier in a boxplot. There is yet another complication here in the use of "hinges" instead of quartiles in finding the $1.5IQR$ values; again, be aware that these values are "suspected" outliers if close.

Chapter 1 Solutions

1.1 Working in seconds means avoiding decimals and fractions.

1.3 Exam1 = 95, Exam2 = 98, Final = 96.

1.5 Cases: apartments. Five variables: rent (quantitative), cable (categorical), pets (categorical), bedrooms (quantitative), distance to campus (quantitative).

1.7 Answers will vary. **(a)** For example, number of graduates could be used for similar-sized colleges. **(b)** One possibility might be to compare graduation rates between private and public colleges.

1.9 (a) The cases are the individual employees. **(b)** The first four (employee identification number, last name, first name, and middle initial) are labels. Department and education level are categorical variables; number of years with the company, salary, and age are quantitative variables. **(c)** Column headings in student spreadsheets will vary, as will sample cases.

1.11 Age: quantitative, possible values 16 to ? (what would the oldest student's age be?). Sing: categorical, yes/no. Can you play: categorical, no, a little, pretty well. Food: quantitative, possible values $0 to ? (what would be the most a person might spend in a week?). Height: quantitative, possible values 2 feet to 9 feet (check the *Guinness Book of World Records*).

1.13 Answers will vary. A few possibilities would be graduation rate, student-professor ratio, and job placement rate.

1.15 Answers will vary. One possibility is alcohol-impaired fatalities per 100,000 residents. This allows comparing states with different populations; however, states with large seasonal populations (like Florida) might be overstated.

1.17 Shown are two possible stemplots; the first uses split stems. The scores are unimodal and slightly left-skewed; most range from 70 to 98. Very few students scored less than 70. There are no apparent outliers.

```
5 | 58          5 | 58
6 | 0           6 | 058
6 | 58          7 | 00235558
7 | 0023        8 | 000135557
7 | 5558        9 | 00023338
8 | 00013
8 | 5557
9 | 0002333
9 | 8
```

1.19 (a) The stemplot is at right. **(b)** Use two stems, even though one is blank. Seeing the gap is useful.

```
1 | 6
2 | 3
2 | 568
3 |
3 | 5555678
4 | 012233
4 | 8
5 | 1
```

1.21 The larger classes hide a lot of detail; there are now only three bars in the histogram.

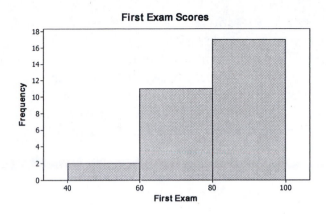

1.23 A stemplot or histogram can be used (possible stemplots are shown in the solution to Exercise 1.17); the distribution is unimodal and left-skewed, centered near 80, and range from 55 to 98. There are no apparent outliers.

1.25 (a) Bar graph on right. **(b)** Second class had the fewest passengers. Third class had by far the most, over twice as many as in first class. **(c)** A bar graph of the percents (relative frequency) would have the same features.

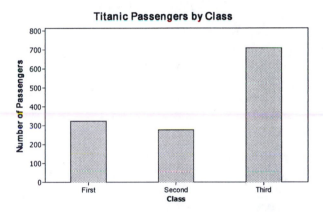

1.27 We divided the number of survivors by the number of passengers in each class, then multiplied by 100 to get a percent. A bar graph is appropriate because we now have three "wholes" to consider.

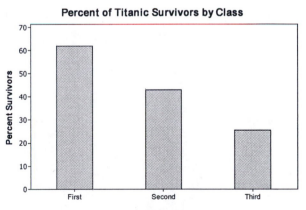

1.29 (a) See histogram on right. **(b)** The overall pattern is unimodal (one major peak). The shape is roughly symmetric with center about 26 and spread from 19 to 33. There appears to be one possible low outlier. **(c)** Answers will vary. The scale on the histogram indicates a possible outlier not seen in the stemplot.

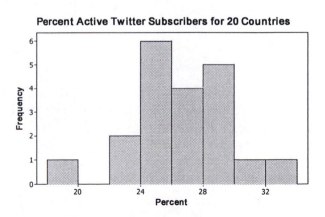

1.31 (a) 2010 still has the highest usage in December and January. See time plot at right. **(b)** The patterns are very similar, but we don't see the increase between February and March that occurred in 2011; consumption in May was slightly higher in 2010. These differences are most likely due to weather.

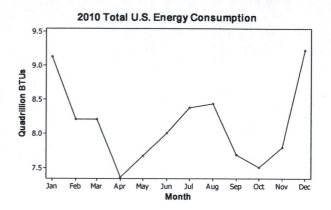

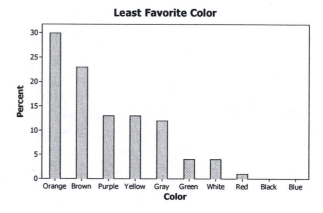

1.33 Bar graph at right. For example, opinions about least-favorite color are somewhat more varied than favorite colors. Interestingly, purple is liked and disliked by about the same percentage of people.

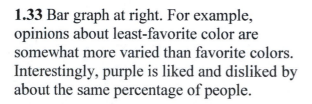

1.35 (a and b) The bar graphs are shown on the next page. **(c)** Preferences will vary, but the ordered bars make it easier to pick out similar categories. The most frequently recycled types (Paper and Trimmings) stand out in both graphs. **(d)** We cannot make a pie chart because each garbage type is a "whole."

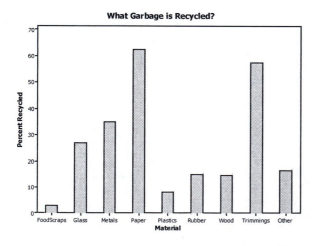

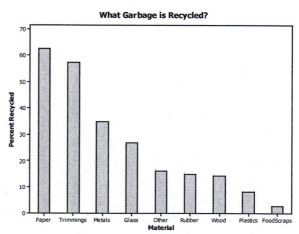

1.37(a and b) The bar graph and pie chart are shown on the next page. Mobile browsers are dominated by Safari (on iPads and iPhones). Android has about one fourth of the market. All others are minor players.

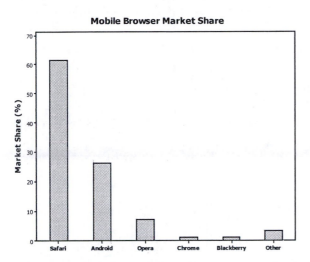

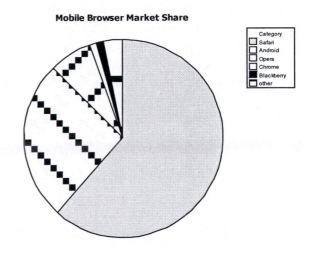

1.39 (a and b) Black is clearly more popular in Europe than in North America. The most popular four colors account for at least 70% of cars in both regions.

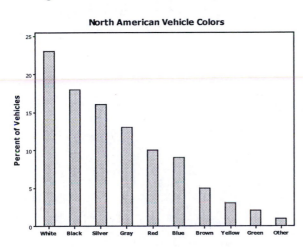

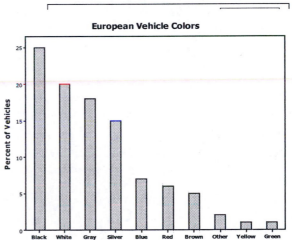

(c) One possibility is shown at right.

1.41 (a)

Region	% FB Users	Region	% FB Users
Africa	3.9	Middle East	9.4
Asia	5.0	North America	49.9
Caribbean	15.4	Oceana Australia	38.9
Central America	26.5	South America	28.1
Europe	28.5		

(b) Answers will vary. When looking only at the absolute number of Facebook users, Europe was the leading region; however, when expressed as a percent of the population, North America has the most Facebook users (almost half of the population). Africa, Asia, and the Middle East have very low percentages of Facebook users. **(c)** The stemplot is shown at right. **(d)** The shape of the distribution might be

```
0 | 359
1 | 5
2 | 688
3 | 8
4 | 9
```

called right-skewed (there are numerical gaps between 28 and 38 and between 38 and 49). The center of the distribution would be about 26% (Central America). This stemplot does not really indicate any major outliers (either high or low). **(e)** Answers will vary, but one possibility is that the scaling in the stemplot actually hides the gaps in the distribution. **(f)** One possibility is that both the population and number of Facebook users are rounded.

1.43 (a) Four variables: GPA, IQ, and self-concept are quantitative; gender is categorical. **(b)** See below, left. **(c)** Unimodal and skewed to the left, centered near 7.8, spread from 0.5 to 10.8. **(d)** There is more variability among the boys; in fact, there seem to be two groups of boys—those with GPAs below 5 and those with GPAs above 5.

```
 0 | 5
 1 | 8
 2 | 4
 3 | 4689
 4 | 0679
 5 | 1259
 6 | 0112249
 7 | 22333556666666788899
 8 | 0000222223347899
 9 | 002223344556668
10 | 01678
```

Female		Male
	0	5
	1	8
	2	4
4	3	689
7	4	069
952	5	1
4210	6	129
98866533	7	223566666789
997320	8	0002222348
65300	9	2223445668
710	10	68

1.45 The stemplot is at right, with split stems. The distribution is unimodal and skewed to the left, with center around 59.5. Most self-concept scores are between 35 and 73, with a few below that, and one high score of 80 (but not really high enough to be an outlier).

```
2 | 01
2 | 8
3 | 0
3 | 5679
4 | 02344
4 | 6799
5 | 1111223344444
5 | 556668899
6 | 00001233344444
6 | 55666677777899
7 | 0000111223
7 |
8 | 0
```

1.47 The new mean is 50.44 (rounds to 50) days.

1.49 The sorted data are

5	5	5	5	6	7	7	7	8	12	12	13	
13	15	18	18	27	28	36	48	52	60	66	94	694

Adding in the outlier adds another observation but does not change the median at all. Instead of being the average of 13 and 13, it is the middle (13th) value, which is 13.

1.51 The sorted data are 55, 73, 75, 80, 81, 85, 90, 93, 93, 98; the median is in position $(10 + 1)/2 = 5.5$, so $M = (81 + 85)/2 = 83$.

1.53 The mean is $\bar{x} = \dfrac{77 + 289 + 128 + \ldots + 25}{80} = 196.575$ minutes (the value 196 in the text was rounded). The quartiles and median are in positions 20.5, 40.5, and 60.5. The sorted data are given below. Based on this, $Q_1 = 54.5$, $M = 103.5$, $Q_3 = 200$.

1	2	2	3	4	9	9	9	11	19
19	25	30	35	40	44	48	51	52	54
55	56	57	59	64	67	68	73	73	75
75	76	76	77	80	88	89	90	102	103
104	106	115	116	118	121	126	128	137	138
140	141	143	148	148	157	178	179	182	199
201	203	211	225	274	277	289	290	325	367
372	386	438	465	479	700	700	951	1148	2631

1.55 Use the five-number summary from Exercise 1.54 (55, 75, 83, 93, 98). The plot is given at right.

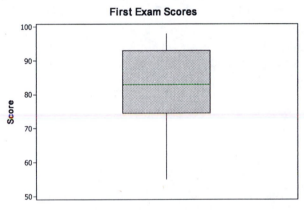

1.57 $s^2 = 159.34$ and $s = 12.62$. To compute this "by hand," recall that we found $\bar{x} = 82.3$ in Exercise 1.48. We would have

$$s^2 = \dfrac{(81 - 82.3)^2 + (73 - 82.3)^2 + \cdots + (90 - 82.3)^2}{9}.$$

1.59 Without Suriname, there are $n = 24$ observations. The median will be located at position 12.5 in the sorted data. Quartiles are located at position 6.5 and 18.5. $Q_1 = 7$ and $Q_3 = 32$. The IQR is 25. Including Suriname, there are now $n = 25$ observations. The median is located at position 13; quartiles will be located at positions 6.5 and 19.5: $Q_1 = 7$ and $Q_3 = 42$. The IQR is now 35. The IQR increased because we now have one more large observation.

1.61 (a) $\bar{x} = 122.9$. **(b)** $M = 102.5$. **(c)** The data set is right-skewed with an outlier (London), so the median is a better center.

1.63 (a) $IQR = 129 - 67 = 62$. **(b)** Outliers are below $67 - 1.5*62 = -26$ or above $129 + 1.5*62 = 222$. London is confirmed as an outlier. **(c)** The boxplots for (c) and (d) are shown on the following page. The first three quarters are about equal in length, and the last (upper quarter) is extremely long. **(d)** The main part of the distribution is relatively symmetric; there is one extreme high outlier. The minimum is about 25, the first quartile is about 70, the median is about 100, and the third quartile is about 125. There is a gap in the data from roughly 200 to about 425. **(e)** The stemplot is at right. **(f)** Answers will vary. For example, the stemplot and the boxplot both indicate the same shape distribution; relatively symmetric with an extremely high outlier. One can just approximate locations of the median and quartiles from the boxplot but get more exact (at least to within rounding) with the stemplot.

```
0 | 24
0 | 6679
1 | 1114
1 | 9
2 |
2 |
3 |
3 |
4 | 2
```

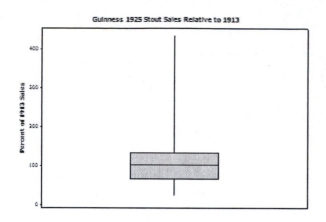

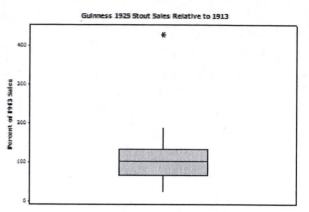

1.65 (a) $s = 8.80$. **(b)** With $n = 50$, the positions of Q_1 and Q_3 will be at 13 and 38 (51 – 13, or 13 in from the upper end of the sorted distribution). We find $Q_1 = 43.79$ and $Q_3 = 57.02$. **(c)** Answers will vary. However, if students said the median was the better center in Exercise 1.64, they should pair that with the quartiles, and not the standard deviation.

1.67 (a) A histogram of the data is at right.
(b) Using software, we find the numerical summaries shown below.

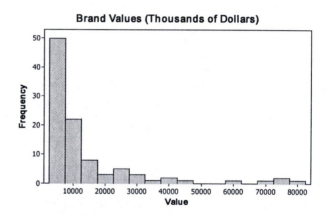

```
   Mean   StDev
  13830   16050


Minimum   Q1   Median     Q3   Maximum
   3731 4775     7516  15537     77839
```
(c) Answers will vary, but due to the severe right-skew, the best measures to describe this distribution are the five-number summary.

1.69 (a) With all the data, the mean is $\bar{x} = 5.23$ and $M = 4.9$. Removing the outliers, we have $\bar{x} = 4.93$ and $M = 4.8$. The mean decreased by 0.3, but M only decreased by 0.1. **(b)** With all the data, $s = 1.429$, $Q_1 = 4.4$, $Q_3 = 5.6$. Removing the outliers, we have $s = 0.818$, $Q_1 = 4.3$, and $Q_3 = 5.5$. The standard deviation decreased by 0.611 percent. Q_1 decreased by 0.1, Q_3 also decreased by 0.1. **(c)** Answers will vary. This distribution is fairly symmetric with a large number of the beers clustered near the mean. The main difference is reducing the standard deviation by eliminating the outliers.

1.71 (a) With a small data set, a stemplot is reasonable. There are clearly two clumps of data. Summary statistics are shown below.

```
Mean    StDev   Minimum     Q1   Median      Q3   Maximum
6.424   1.400       3.7   4.95      6.7    7.85         8
```
(b) Because of the clusters of data, one set of numerical summaries will not be adequate. **(c)** After separating the data, we have for the smaller weights

```
  Mean    StDev   Minimum     Q1   Median      Q3   Maximum
 4.662    0.501       3.7    4.4      4.7   5.075       5.3
```

```
3 | 7
4 | 3
4 | 7777
5 | 23
5 |
6 | 0033
6 | 79
7 | 0
7 | 666899999
8 | 0
```

And for the larger weights

Mean	StDev	Minimum	Q1	Median	Q3	Maximum
7.253	0.740	6	6.5	7.6	7.9	8

1.73 (a) 0, 0, 5.09, 9.47, 73.2. **(b and c)** See graphs on the next page. **(d)** Answers will vary. The distribution is strongly skewed right with five high outliers.

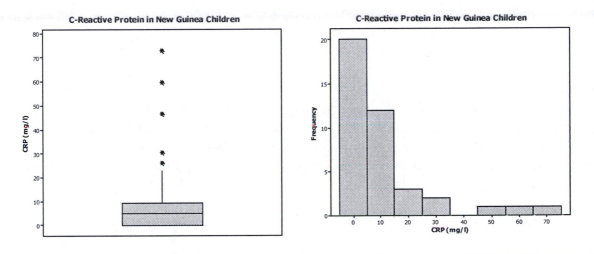

1.75 This distribution is unimodal and skewed right and has no outliers, as shown in the graphs following. The five-number summary is 0.24, 0.355, 0.76, 1.03, 1.9.

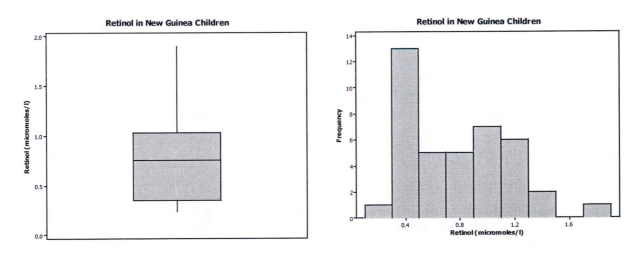

1.77 Some people like celebrities and business executives make a very large amount of money and have large-valued assets (think Bill Gates of Microsoft, Warren Buffett, Oprah, etc.).

1.79 The mean is $\mu = \dfrac{6 * 45000 + 2 * 70000 + 420000}{9} = \$92,222.22$. Eight of the employees make less than this. $M = \$45,000$.

1.81 The median doesn't change, but the mean increases to $101,111.11.

1.83 The average would be 2.5 or less (an earthquake that isn't usually felt). These do little or no damage.

1.85 For $n = 2$ the median is also the average of the two values.

1.87 (a) There are several different answers, depending on the configuration of the first five points. *Most students* will likely assume that the first five points should be distinct (no repeats), in which case the sixth point *must* be placed at the median. This is because the median of five (sorted) points is the third, while the median of six points is the average of the third and fourth. If these are to be the same, the third and fourth points of the set of six must both equal the third point of the set of five.

   The diagram below illustrates all of the possibilities; in each case, the arrow shows the location of the median of the initial five points, and the shaded region (or dot) on the line indicates where the sixth point can be placed without changing the median. Notice that there are four cases where the median does not change, regardless of the location of the sixth point. (The points need not be equally spaced; these diagrams were drawn that way for convenience.)

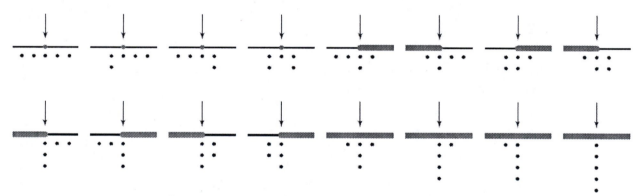

(b) Regardless of the configuration of the first five points, if the sixth point is added so as to leave the median unchanged, then in that (sorted) set of six, the third and fourth points must be equal. One of these two points will be the middle (fourth) point of the (sorted) set of seven, no matter where the seventh point is placed.

1.89 (a) *Bihai*: $\bar{x} = 47.5975$, $s = 1.2129$. Red: $\bar{x} = 39.7113$, $s = 1.7988$. Yellow: $\bar{x} = 36.1800$, $s = 0.9753$ (all in mm). **(b)** *Bihai* and red appear to be right-skewed (although it is difficult to tell with such small samples). Skewness would make these distributions unsuitable for $\bar{x}$ and s.

Bihai		Red		Yellow	
46	3466789	37	4789	34	56
47	114	38	0012278	35	146
48	0133	39	167	36	0015678
49		40	56	37	01
50	12	41	4699	38	1
		42	01		
		43	0		

1.91 Answers will vary. Take six or more numbers, with the smallest number much smaller than Q_1. For example, 1, 2, 10, 11, 12, 13 has $\bar{x} = 8.17$. The five-number summary is 1, 2, 10.5, 12, 13.

1.93 (a) Answers will vary. Any set of four identical numbers works. **(b)** 0, 0, 20, 20 is the only possible answer.

1.95 Answers will vary. Typical calculators will carry only about 12 to 15 digits; for example, a TI-84 fails (gives $s = 0$) for 14-digit numbers. *Excel* (at least the version checked by the author) also fails for 16-digit numbers and gave $s = 0$ (checking further, all the values were stored as 3000000000000000).

1.97 $\bar{x} = 2.32*2.2 = 5.104$ pounds and $s = 1.21*2.2 = 2.662$ pounds.

1.99 Full data set: $\bar{x} = 196.575$ and $M = 103.5$ minutes. The 10% and 20% trimmed means are $\bar{x} = 127.734$ and $\bar{x} = 111.917$ minutes. While still larger than the median of the original data set, they are much closer to the median than the ordinary untrimmed mean.

1.101 The range is from $288 - 2*38 = 212$ to $288 + 2*38 = 364$.

1.103 $z = \dfrac{365 - 288}{38} = 2.03$.

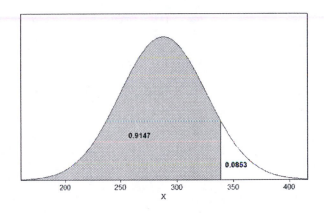

1.105 $z = \dfrac{340 - 288}{38} = 1.37$. Using Table A, the proportion below 340 is 0.9147, and the proportion at or above is 0.0853. Using technology, the proportion below 340 is 0.9144.

1.107 $x = \mu + z\sigma$. From Table A, we find the area to the left of $z = 0.67$ to be 0.7486 and the area to the left of $z = 0.68$ to be 0.7517. (Technology gives $z = 0.6745$.) If we approximate as $z = 0.675$, we have $x = 288 + 0.675*38 = 313.65$, or about 314. If you use $z = 0.67$, the answer would be 313.46.

1.109 (a) The mean and median are both 1.

(b) The mean is about -1 and the median about -0.3.

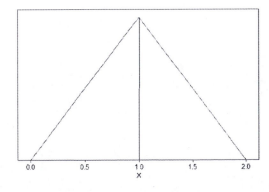

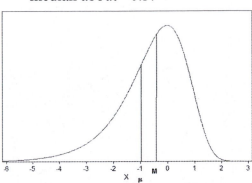

1.111 (a and b) The distributions are shown at right. **(c)** The distributions look the same, only shifted.

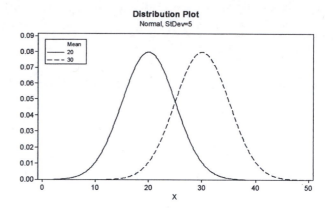

1.113 (a) The distribution is shown at right. **(b and c)** The table below indicates the desired ranges. These values are shown on the *x* axis of the distribution.

	Low	High
68%	256	320
95%	224	352
99.7%	192	384

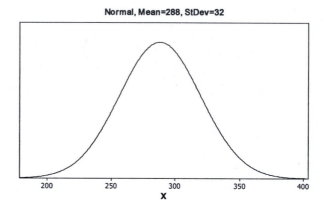

1.115

Value	Percentile (Table A)	Percentile (Software)
150	50	50
140	38.6	38.8
100	7.6	7.7
180	80.5	80.4
230	98.9	98.9

1.117 Using the $N(153,34)$ distribution, we find the values corresponding to the given percentiles as given below (using Table A). The actual scores are very close to the percentiles of the Normal distribution; we can conclude these scores are at least approximately Normal.

Percentile	Score	Score with $N(153, 34)$
10%	110	109
25%	130	130
50%	154	153
75%	177	176
90%	197	197

1.119 (a) Ranges are given in the table at right. In both cases, some of the lower limits are negative, which does not make sense; this happens because the women's distribution is skewed, and the men's distribution has an outlier. Contrary to the conventional wisdom, the men's

	Women	Men
68%	8489 to 20,919	7158 to 22,886
95%	2274 to 27,134	−706 to 30,750
99.7%	−3941 to 33,349	−8570 to 38,614

mean is slightly higher, although the outlier is at least partly responsible for that. **(b)** The means suggest that Mexican men and women tend to speak more than people of the same gender from the United States.

1.121 (a) F: −1.645. D: −1.04. C: 0.13. B: 1.04. **(b)** F: below 55.55. D: between 55.55 and 61.6. C: between 61.6 and 73.3. B: between 73.3 and 82.4. A: above 82.4. **(c)** Opinions will vary.

1.123 (a) 1/5 = 0.2. **(b)** 1/5 = 0.2. **(c)** 2/5 = 0.4.

1.125 (a) Mean is C, median is B (the right-skew pulls the mean to the right). **(b)** Mean A, median A. **(c)** Mean A, median B (the left-skew pulls the mean to the left).

1.127. (a) The applet shows an area of 0.6826 between −1.000 and 1.000, while the 68–95–99.7 rule rounds this to 0.68. **(b)** Between −2.000 and 2.000, the applet reports 0.9544 (compared to the rounded 0.95 from the 68–95–99.7 rule). Between −3.000 and 3.000, the applet reports 0.9974 (compared to the rounded 0.997).

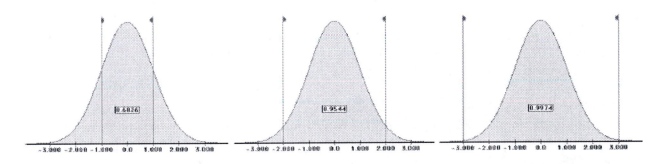

1.129

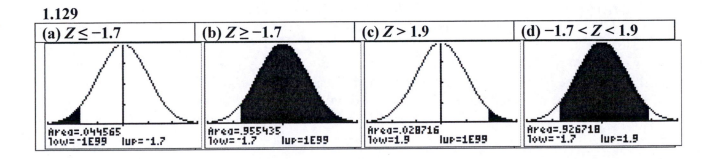

(a) $Z \le -1.7$	(b) $Z \ge -1.7$	(c) $Z > 1.9$	(d) $-1.7 < Z < 1.9$
Area=.044565 low=-1E99 lup=-1.7	Area=.955435 low=-1.7 lup=1E99	Area=.028716 low=1.9 lup=1E99	Area=.926718 low=-1.7 lup=1.9

1.131

(a) Cumulative 0.78, $z = 0.77$	(b) $Z > z$ has proportion 0.22, $z = 0.77$
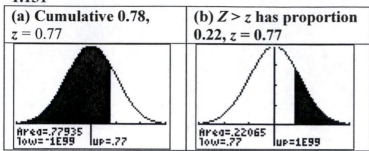	

1.133 130 is a z-score of 2. From Table A, 2.28% of people would qualify for MENSA membership.

1.135 Joshua's z-score is $(17 - 21.5)/5.4 = -0.83$. Anthony has a z-score of $(1030 - 1498)/316 = -1.48$. Joshua's score is higher.

1.137 Alyssa's z-score is $(32 - 21.5)/5.4 = 1.94$. The equivalent SAT score would be $1498 + 1.94*316 = 2111.04$, or about 2111.

1.139 Joshua's z-score is $(17 - 21.5)/5.4 = -0.83$. Table A gives 0.2033, so he is at the 20th percentile.

1.141 From Table A, $z = -1.28$ is the closest value corresponding to 10% below z. We have Score $= 1498 - 1.28*316 = 1093.52$, so about 1094 and lower.

1.143 From Table A, the quartiles have z-scores of -0.675, 0, and 0.675. This gives scores of 1285, 1498, and 1711 (rounded to the nearest integer).

1.145 (a) $z = (40 - 46)/13.6 = -0.44$. From Table A, 33% of men have low values of HDL. (Software gives 32.95%.) **(b)** $z = (60 - 46)/13.6 = 1.03$. From Table A, 15.15% of men have protective levels of HDL. (Software gives 15.16%.) **(c)** $(1 - 0.1515) - 0.33 = 0.5185$. 51.85% of men are in the intermediate range for HDL. (Software gives 0.5188.)

1.147 (a) The first and last deciles for a standard Normal distribution are ± 1.2816. **(b)** For a $N(9.12, 0.15)$ distribution, the first and last deciles are $\mu - 1.2816\sigma = 8.93$ and $\mu + 1.2816\sigma = 9.31$ ounces.

1.149. (a) As the quartiles for a standard Normal distribution are ± 0.6745, we have $IQR = 1.3490$. **(b)** $c = 1.3490$: For a $N(\mu, \sigma)$ distribution, the quartiles are $Q_1 = \mu - 0.6745\sigma$ and $Q_3 = \mu + 0.6745\sigma$, so $IQR = (\mu + 0.6745\sigma) - (\mu - 0.6745\sigma) = 1.3490\sigma$.

1.151 To find these levels, use $55 + z*15.5$, where z from Table A has the given percent as the area to the left of z. With symmetry, $z_{10\%} = -z_{90\%}$.

Percentile	10%	20%	30%	40%	50%	60%	70%	80%	90%
HDL level	35.2	42.0	46.9	51.1	55	58.9	63.1	68.0	74.8

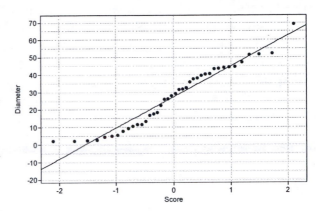

1.153 (a) The yellow variety is the nearest to a straight line. **(b)** The other two distributions are both slightly right-skewed, and the *bihai* variety appears to have a couple of high outliers. **(c)** The deviations do not appear to be approximately Normal. There are too many lengths at the low end that are similar (the horizontal "tail") and a jump and then gapping at the high end (outliers?). Putting these two observations together, the distribution appears to be right-skewed.

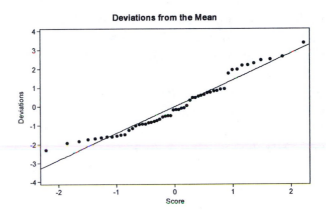

Deviations from the Mean

1.155 Answers will vary. The histogram suggests Figure 1.32, but is not perfectly level across the top. The uniform distribution does not extend as low or as high as a Normal distribution, so on each end of the distribution we have several observations with essentially the same values. This causes the bending of the line seen in the Normal quantile plot.

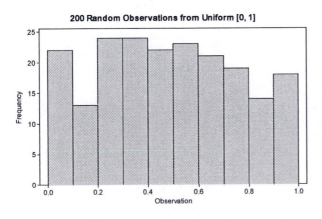

200 Random Observations from Uniform [0, 1]

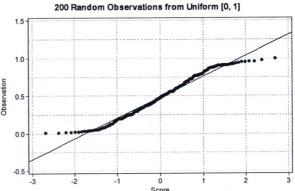

200 Random Observations from Uniform [0, 1]

1.157 (a) The distribution appears to be roughly Normal, apart from two possible low and two possible high outliers. **(b)** The outliers on either end would inflate the standard deviation. The five-number summary is 8.5, 13.15, 15.4, 17.8, 23.8. **(c)** For example, smoking rates are typically 12% to 20%. Which states are high, and which are low?

```
 8 | 56
 9 |
10 | 8
11 | 48
12 | 0112348
13 | 123557
14 | 3344579
15 | 0455699
16 | 34699
17 | 356
18 | 0368
19 | 13449
20 | 1
21 | 1
22 |
23 | 68
```

1.159 Students might compare color preferences using a stacked bar graph like that shown on the right. (They could also make six pie charts, but comparing slices across pies is difficult.) Possible observations: white is least popular in China, and silver is less common in Europe. Silver, white, gray, and black dominate the market worldwide.

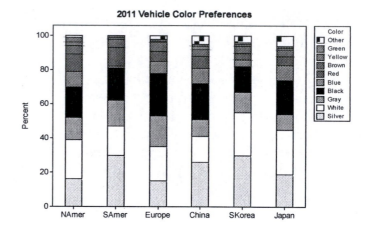

1.161 Answers will vary.

1.163 (a) This distribution is more decidedly right-skewed than the 2011 distribution. We can note that number of countries in the left-most two bars (less than 20 Internet users per 100 people) declined from 2010 to 2011. Because of the distribution shape, we compute the five-number summary.

```
Min      Q1    Median     Q3     Max
0.21   10.31    31.40   55.65   95.63
```

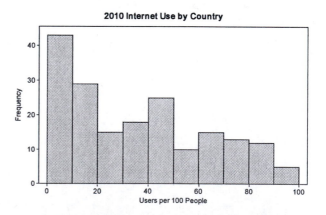

(b) The change in Internet users is definitely right-skewed. The Slovak Republic actually declined by 1.28 users per 100 people; Rwanda also decreased (from 8 to 7 users per 100 people). There were a few countries with miniscule declines. Two countries (the Bahamas and Bahrain) increased by 22 users per 100 people. The five-number summary follows.

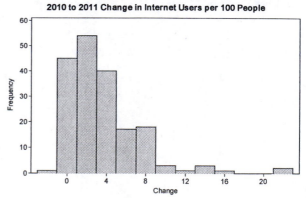

```
   Min      Q1     Med      Q3      Max
-1.285   0.996   2.570   4.811   22.000
```

(c) Percent change in Internet users per 100 people is extremely right-skewed. Two countries (Myanmar and Timor-Leste) effectively tripled the number of Internet users per 100 people, but they still have less than 1 Internet user per 100 people.

```
   Min      Q1     Med      Q3      Max
-12.50    5.58   10.75   20.70   327.32
```

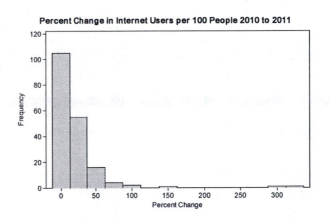

1.165 AT&T and Comcast are by far the largest Internet providers. Some of the smaller providers (Earthlink and Suddenlink, for example) could be taken over.

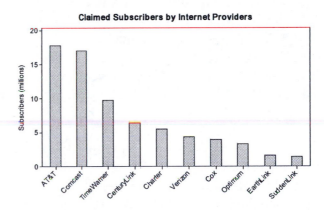

1.167 (a) For car makes (a categorical variable), use either a bar graph or pie chart. For car age (a quantitative variable), use a histogram, stemplot, or boxplot. **(b)** Study time is quantitative, so use a histogram, stemplot, or boxplot. To show change over time, use a time plot (average hours studied against time). **(c)** Use a bar graph or pie chart to show radio station preferences. **(d)** Use a Normal quantile plot to see whether the measurements follow a Normal distribution.

1.169 (a) RDA = 75 = 60 + $z\sigma$. From Table A, $z = 2.00$. We have $15 = 2.00\sigma$, so $\sigma = 7.5$. **(b)** The distribution is shown at right. Note that the UL is far off into the upper tail at 2000 mg/d.

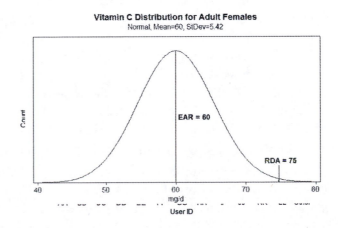

1.171 (a) We were given that the mean is 84.1 and the 50th percentile was 79. In a Normal distribution, these should be equal; one option would be to average these, and estimate $\mu = \dfrac{79 + 84.1}{2} = 81.55$. The 5th percentile is $42 = 81.55 - 1.645\sigma$. This implies $\sigma = 24.04$. The 95th percentile is $142 = 81.55 + 1.645\sigma$. This implies $\sigma = 36.75$. If we average the two estimates, we would have $\sigma = 30.4$. **(b)** The graph is at right. **(c)** From the two distributions, over half of women consume more vitamin C than they need, but some consume far less.

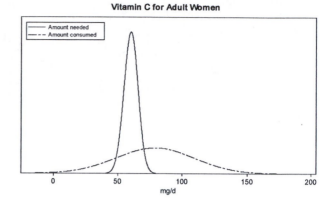

1.173 (a) Not only are most responses multiples of 10, but many are multiples of 30 and 60. Most people will "round" their answers when asked to give an estimate like this; in fact, the most striking answers are ones such as 115, 170, or 230. The students who claimed 360 minutes (6 hours) and 300 minutes (5 hours) may have been exaggerating. **(b)** Women seem to generally study more (or claim to), as there are none that claim less than 60 minutes per night. The center (median) for women is 170; for men the median is 120 minutes. **(c)** The boxplots are given. Opinions will vary.

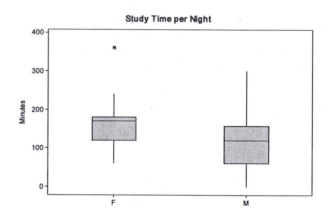

1.175 No, and no: It is easy to imagine examples of many different data sets with mean 0 and standard deviation 1, for example, $\{-1, 0, 1\}$ and $\{-2, 0, 0, 0, 0, 0, 0, 0, 2\}$. Likewise, for any given five numbers $a \le b \le c \le d \le e$ (not all the same), we can create many data sets with that five-number summary, simply by taking those five numbers and adding some additional numbers in between them, for example (in increasing order): 10, , 20, , , 30, , , 40, , 50. As long as the number in the first blank is between 10 and 20, and so on, the five-number summary will be 10, 20, 30, 40, 50.

1.177 Results will vary. One set of 20 samples gave the results at the right (Normal quantile plots are not shown). Theoretically, $\bar{x}$ will have a Normal distribution with mean 25 and standard deviation $8/\sqrt{30} \doteq 1.46$, so that about 99.7% of the time, one should find $\bar{x}$ between 20.6 and 29.4. Meanwhile, the theoretical distribution of s is nearly Normal (slightly skewed) with mean $= 7.9313$ and standard deviation $= 1.0458$; about 99.7% of the time, s will be between 4.8 and 11.1.

Means		Standard Deviations	
		5	3
		5	
22	9	6	
23	689	6	8
24	019	7	023334
25	138	7	78
26	0036689	8	002233
27	247	8	7
		9	01
		9	8

Chapter 2 Solutions

2.1 (a) The 30 students. **(b)** Attendance and score on the final exam. **(c)** Score on the final is quantitative. Attendance is most likely quantitative: number of classes attended (or missed).

2.3 Cases: cups of Mocha Frappuccino. Variables: size and price (both quantitative).

2.5 (a) Tweets. **(b)** Click count and length of tweet are quantitative. Day of week and gender are categorical. Time of day could be quantitative (as hh:mm) or categorical (if morning, afternoon, etc.). **(c)** Click counts is the response. The others could all be potentially explanatory.

2.7 Answers will vary. Some possible variables are condition, number of pages, and binding type (hardback or paperbound), in addition to purchase price and buyback price. Cases would be individual textbooks. Here, we would likely be interested in predicting the difference in purchase and buyback price based on the other variables.

2.9 (a) Temperatures are usually similar from one day to the next (recording temperatures at noon each day, for example). One variable that would help is whether a front (cold or warm) came through. The individuals would be the days. **(b)** No relationship. These are different individuals. **(c)** Answers will vary. It's possible quality and price are related, but not certain.

2.11 Histograms of both variables are below. Summary statistics are below. Price per load looks right-skewed. Quality rating has two different clusters of values.

```
Variable        Mean   StDev  Minimum    Q1  Median     Q3  Maximum
Rating         43.88   10.77    26.00  33.50  47.00  51.50    61.00
PricePerLoad   14.21    5.99     5.00  10.00  13.50  17.00    30.00
```

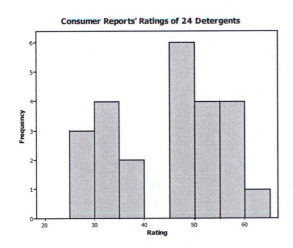

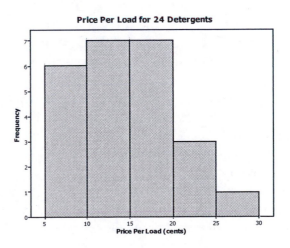

2.13 (b) The graph is at right. **(c)** The only difference is in the scaling of the *x* axis.

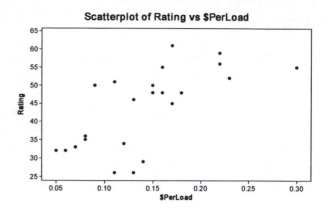

2.15 Create a new variable such as the ratio of the two (Debt2010/Debt2009). If the variables are approximately the same, this should have mean 1 and a small standard deviation. The mean of this distribution is $\bar{x} = 1.082$ and the standard deviation is $s = 0.1427$. A stemplot of the new variable is at right.

```
 8  | 8
 9  | 022
 9  | 67788
10  | 00111244
10  | 57
11  | 00112234
11  | 56
12  |
12  | 8
13  | 3
13  |
14  |
14  | 78
```

2.17 (a) All the liquid detergents are at the upper right and the powder detergents at the lower left, so type of detergent seems to highly influence the relationship. **(b)** Students should answer that liquids cost more and have higher ratings; reasons will vary.

2.19 (a) The scatterplot is at right. **(b)** The overall pattern is linear and increasing. There is possibly one outlier at the upper right, far from the other points. **(c)** The relationship is roughly linear, increasing, and moderately strong. **(d)** The baseball player represented by the point at the far right is not as strong in his dominant arm as other players. **(e)** Other than the one outlier, the relationship is approximately linear.

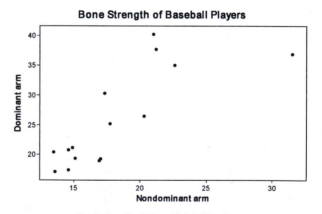

2.21 (a) Population should be the explanatory variable and college students the response. **(b)** See plot at right. This graph shows a strong, linear, increasing relationship with one high outlier both in undergraduates and population (California).

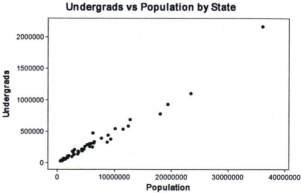

2.23 (a) See plot at right. **(b and c)** The relationship is very strong, linear, and decreasing. **(d)** There do not appear to be any outliers. **(e)** The relationship is linear.

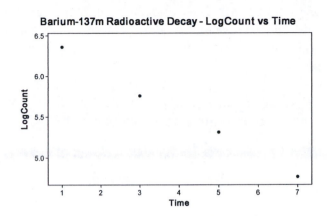

2.25 (a) The description is for variables that are positively related. **(b)** The response variable is plotted on the *y* axis, and the explanatory variable on the *x* axis. **(c)** A histogram shows the distribution of a single variable, not the relationship between two variables. A scatterplot can be used to examine the relationship between two quantitative variables.

2.27 (a) The scatterplot is at right. **(b)** The relationship is linear, increasing and much stronger than the relationship between carbohydrates and percent alcohol. Most of the scatter is relatively close to the center of the plot. The points at the upper right could be considered to be outliers (especially the two that do not really follow the linear pattern.

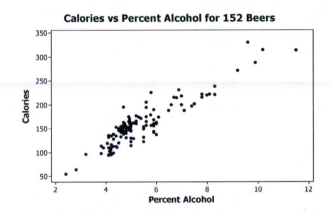

2.29 (a) The scatterplot is at right. **(b)** This plot is much more linear than the original data scatterplot. **(c)** Answers will vary (students perhaps won't like using logs), but in terms of regression analysis, the log data could easily have a line fit to them.

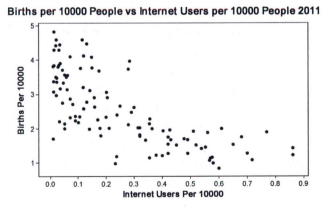

2.31 (a) Examine for a relationship. **(b)** Use high school GPA as explanatory and college GPA as response. **(c)** Use square feet as explanatory and rental price as response. **(d)** Use amount of sugar as explanatory and sweetness as response. **(e)** Use temperature yesterday at noon as explanatory and temperature today at noon as response.

2.33 (a) In general, we expect more intelligent children to be better readers, and less intelligent children to be weaker readers. The plot does show this positive association. **(b)** These four have moderate IQs but poor reading scores. **(c)** The relationship is roughly linear but somewhat weak (much scatter).

2.35 (a) The scatterplot is at right. **(b)** The association is positive and linear. Overall the relationship is strong, but it is stronger for women than for men. Male subjects generally have both larger lean body mass and higher metabolic rates than women.

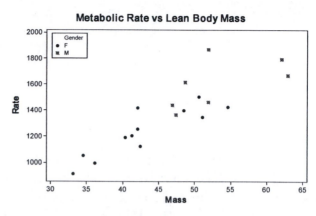

2.37 (a) Both show fairly steady improvement. Women have made more rapid progress, but have not improved since 1993, whereas men's records may be dropping more rapidly in recent years. **(b)** The data support the first claim but do not seem to support the second; there is no information about sprints here.

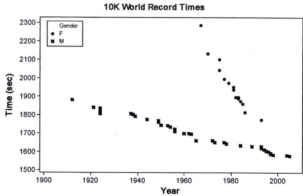

2.39 (a) This is a linear transformation. Dollars = 0 + 0.01*Cents. **(b)** $r = 0.671$. **(c)** They are the same. **(d)** Changing units does not change r.

2.41 (a) No *linear* relationship. (There could be a *nonlinear* relationship, though.) **(b)** Strongly linear and negative. **(c)** Weakly linear and positive. **(d)** Strongly linear and positive.

2.43 (a) $r = 0.905$. **(b)** The plot is at right. Correlation is a good summary for these data. The pattern is linear and appears to be strong. There is, however, one outlier at the upper right.

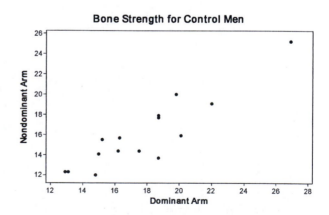

2.45 (a) $r = 0.984$. **(b)** The correlation may be a good summary for these data because the scatterplot is strongly linear. California, however, is an outlier that strengthens the relationship (makes r closer to 1). **(c)** Eliminate California, Texas, Florida, and New York. $r = 0.971$. Expanding the range of values can strengthen a relationship (if the new points follow the rest of the data).

2.47 (a) $r = -0.999$. **(b)** Correlation is a good numerical summary here because the scatterplot is very strongly linear. **(c)** You must be careful; there can be a strong correlation between two variables even when the relationship is curved. Plot the data first!

2.49 (a) r would be 1. (The relationship would be exactly StoreBrand $= 0 + 0.8*$BrandName.) **(b)** r would be 1. (The relationship would be exactly StoreBrand $= -2 +$ BrandName.)

2.51 $r = 0.521$.

2.53 (a) The plot is at right. $r = -0.730$. **(b)** The relationship is curved; birth rate declines with increasing Internet use until about 40 Internet users per 100 people. After that, there is a steady overall birth rate. Correlation is not a good numerical summary for this relationship.

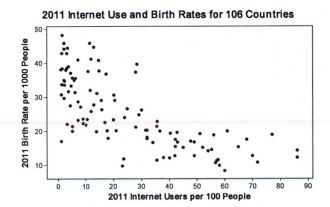

2.55. (a) As two points determine a line, the correlation is always either -1 or $+1$. **(b)** Sketches will vary; an example is shown as the first graph following. Note that the scatterplot must be positively sloped, but r is affected only by the scatter about a line drawn through the data points, not by the steepness of the slope. **(c)** The first nine points cannot be spread from the top to the bottom of the graph because in such a case the correlation cannot exceed about 0.66 (based on empirical evidence—that is, from a reasonable amount of playing around with the applet). One possibility is shown as the second graph below. **(d)** To have $r = 0.8$, the curve must be higher at the right than at the left. One possibility is shown as the third graph below.

2.57 The correlation is $r = 0.481$. The correlation is greatly lowered by the one outlier. Outliers tend to have fairly strong effects on correlation; it is even stronger here because there are so few observations.

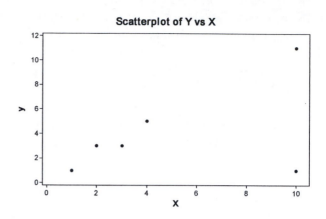

2.59. The person who wrote the article interpreted a correlation close to 0 as if it were a correlation close to −1 (implying a negative association between teaching ability and research productivity). Professor McDaniel's findings mean there is little linear association between research and teaching—for example, knowing that a professor is a good researcher gives little information about whether she is a good or bad teacher.

   Note: *Students often think that "negative association" and "no association" mean the same thing. This exercise provides a good illustration of the difference between these terms.*

2.61. Both relationships (scatterplots follow) are somewhat linear. The GPA/IQ scatterplot ($r = 0.634$) shows a stronger association than GPA/self-concept ($r = 0.542$). The two students with the lowest GPAs stand out in both plots; a few others stand out in at least one plot. Generally speaking, removing these points raises r (because the remaining points look more linear). An exception: Removing the lower-left point in the self-concept plot decreases r because the relative scatter of the remaining points is greater.

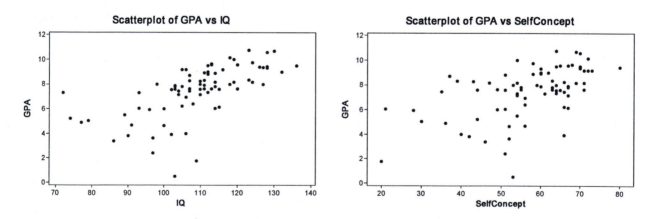

2.63 We predict fat gain $= 3.505 - 0.00344 * 500 = 1.785$ kilograms gained.

2.65

r	−0.8	−0.4	−0.2	0	0.3	0.5	0.9
% variation explained (r^2)	64	16	4	0	9	25	81

From the table above, the percent of variation explained by a regression is always smaller than $|r|$ and drops off drastically as $|r|$ moves away from 1; because squared quantities are always positive, this cannot tell us the direction of the relationship.

2.67 (a) The scatterplot with the least-squares line is shown. **(c)** Descriptions will vary, but the relationship is roughly linear. Bone strength in the dominant arm increases about 1.373 units for every unit increase in strength in the nondominant arm. This increase is more than in the controls.

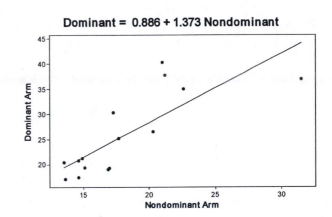

2.69 Predicted bone strength is $0.886 + 1.373*16 = 22.854$ cm^4/1000.

2.71

		Count = 602.8 − (74.7 × time)			**(d)** Count = 500 − (100 × time)		
Time	Count	Predicted **(a)**	Difference **(b)**	Squared Difference **(c)**	Predicted	Difference	Squared Difference
1	578	528.1	49.9	2490.01	400	178	31,684
3	317	378.7	−61.7	3806.89	200	117	13,689
5	203	229.3	−26.3	691.69	0	203	41,209
7	118	79.9	38.1	1451.61	−200	318	101,124

(e) In terms of least squares, the first line is a better description of the relationship between time and radioactive counts. The sum of its squared differences (the least squares that is minimized in regression) is 8440.2; the sum for the second line is 187,706.

2.73 The slope is $b = r\dfrac{s_y}{s_x} = 0.98367\dfrac{358,460}{6,620,733} = 0.0533$. The intercept is $a = \bar{y} - b\bar{x} = 302,136 - (0.0533)(5,955,551) = -15,294.8683$. The estimated regression line is $\widehat{\text{Students}} = -15,294.868 + 0.0533\text{Population}$.

2.75 (a) $\widehat{\text{Students}} = -15,294.868 + 0.0533(6,000,000) = 304,505.132$. **(b)** $\widehat{\text{Students}} = 8487.47 + 0.04846(6,000,000) = 299,247.47$. **(c)** The estimate that included the four largest states (who also have the largest number of undergraduates) is larger than the estimate that excludes them by about 5000 undergraduates.

2.77 (a) $\widehat{\text{Students}}$ = 8491.907 + 0.048Population. **(b)** r^2 = 0.942. **(c)** About 94.2% of the variability in number of undergraduates is explained by the regression on population. **(d)** The numerical output does not tell us whether the relation is linear.

2.79 (a) $\widehat{\text{Carbs}}$ = 2.606 + 1.789PercentAlcohol **(b)** r^2 = 0.271.

   Note: *As we would guess from the scatterplot, and from r^2 = 27.1%, this is not a very reliable prediction.*

2.81 (a) To three decimal places, the correlations are all approximately 0.816 (for set D, r actually rounds to 0.817), and the regression lines are all approximately $\hat{y}$ = 3.000 + 0.500x. For all four sets, we predict $\hat{y}$ = 8 when x = 10. **(b)** Scatterplots below. **(c)** For Set A, the use of the regression line seems to be reasonable—the data do seem to have a moderate linear association (albeit with a fair amount of scatter). For Set B, there is an obvious *non*-linear relationship; we should fit a parabola or other curve. For Set C, the point (13, 12.74) deviates from the (highly linear) pattern of the other points; if we can exclude it, the (new) regression formula would be very useful for prediction. For Set D, the data point with x = 19 is a very influential point—the other points alone give no indication of slope for the line. Seeing how widely scattered the y coordinates of the other points are, we cannot place too much faith in the y coordinate of the influential point; thus, we cannot depend on the slope of the line, so we cannot depend on the estimate when x = 10. (We also have no evidence as to whether or not a line is an appropriate model for this relationship.)

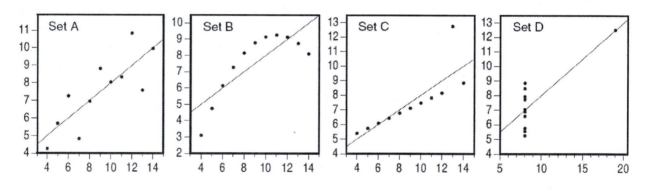

2.83 (a) The added point is an outlier that does not follow the pattern of the rest of the data. It is an outlier in the x direction, but not in y. **(b)** The new regression equation is $\hat{y}$ = 27.56 + 0.1031x. **(c)** r^2 = 0.052. This added point is influential both to the regression equation (the intercept and slope both changed substantially from $\hat{y}$ = 17.38 + 0.6233x) as well as correlation.

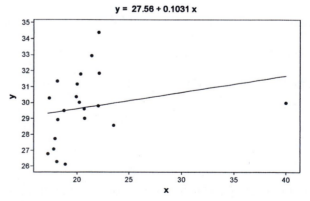

2.85 (a) $y = 12 + 8(3) = 36$. **(b)** y increases by 8 (the value of the slope) when x increases one unit. **(c)** The intercept is 12 (the value of y when $x = 0$).

2.87 The correlation of IQ with GPA is $r_1 = 0.634$; for self-concept and GPA, $r_2 = 0.542$. IQ does a slightly better job; it explains about $r^2 \doteq 40.2\%$ of the variation in GPA, whereas self-concept explains about $r^2 = 29.4\%$ of the variation.

2.89 The slope is $b_1 = r\dfrac{s_y}{s_x}$, and the intercept is $b_0 = \bar{y} - b_1\bar{x}$. The equation of the line is $y = b_0 + b_1x$. When $x = \bar{x}$, $y = (\bar{y} - b_1\bar{x}) + b_1\bar{x} = \bar{y}$.

2.91 For scatterplots and correlation, see the solutions to Exercises 2.36 and 2.54. The regression equations are $\widehat{\text{Value}} = 1073.87 + 1.74\,\text{Debt}$ ($r^2 = 0.5\%$), $\widehat{\text{Value}} = -262.4 + 4.966\,\text{Revenue}$ ($r^2 = 92.7\%$), and $\widehat{\text{Value}} = 872.6 + 5.695\,\text{Income}$ ($r^2 = 79.4\%$). Student reports will vary, but should cite the strong, positive nature of the relationship between team value and revenue; they should also note that the relationship between value and debt is much weaker.

2.93 The residuals sum to 0.01.

2.95 The residuals are calculated as Dominant $- (0.886 + 1.373 \times \text{nondominant})$.

ID	Dominant	Nondominant	Residual	ID	Dominant	Nondominant	Residual
1	19.3	17.0	−4.93	3	25.2	17.7	0.01
2	19.0	16.9	−5.09	4	37.7	21.1	7.71

2.97 (a and b)

Time	LogCount	Predicted	Residual
1	6.35957	6.332444	0.027126
3	5.75890	5.811208	−0.052310
5	5.31321	5.289972	0.023238
7	4.77068	4.768736	0.001944

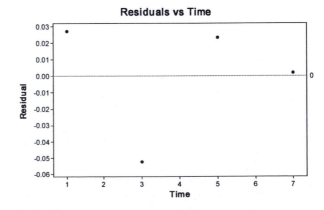

Residuals vs Time

(c) This residuals plot looks random, so this regression is a better model than the one in Exercise 2.96.

2.99 (a and b) Plots are below. **(c, d, and e)** California does not look like an outlier in either distribution, nor in terms of the relationship. **(f)** Removing California changes the regression equation from $\widehat{\text{LogUndergrads}} = -2.574 + 0.9731\,\text{LogPopulation}$ ($r^2 = 97.5\%$) to $\widehat{\text{LogUndergrads}} = -2.40 + 0.9614\,\text{LogPopulation}$ ($r^2 = 97.3\%$). California is not influential to either correlation or the regression in the log scale.

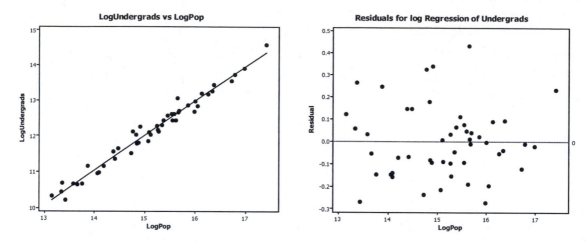

2.101 (a and b) Scatterplots are below. The regression equation is $\overline{\text{Log(UGrad2011)}}$ = 0.007 + 1.01 Log(UGrad2006). **(c)** One data point stands out in both graphs, it is West Virginia with the largest positive residual. The next largest positive residual belongs to Iowa. These do not seem to be influential. **(d and e)** Using the log data removes California as a potentially influential outlier. The data are more equally spread across the range. One possible disadvantage of using the log data is that explaining this to people could be difficult.

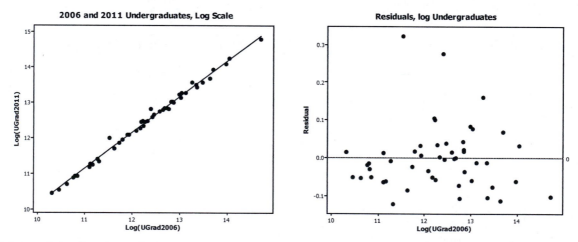

2.103 (a) If the line is pulled toward the influential point, the observation will not necessarily have a large residual. **(b)** High correlation is always present if there is causation. **(c)** Extrapolation is using a regression to predict for *x*-values outside the range of the data (here, using 20, for example).

2.105 Internet use does not cause people to have fewer babies. Possible lurking variables are economic status of the country, levels of education, ...

2.107 Answers will vary. For example, a reasonable explanation is that the cause-and-effect relationship goes in the other direction: Doing well makes students or workers feel good about themselves, rather than vice versa.

2.109 The explanatory and response variables were "consumption of herbal tea" and "cheerfulness/health." The most important lurking variable is social interaction; many of the nursing home residents may have been lonely before the students started visiting.

2.111 (a) Drawing the "best line" by eye is a very inaccurate process; few people choose the best line (although you can get better at it with practice). **(b)** Most people tend to overestimate the slope for a scatterplot with $r \doteq 0.7$; that is, most students will find that the least-squares line (the one without the ending dots) is less steep than the one they draw.

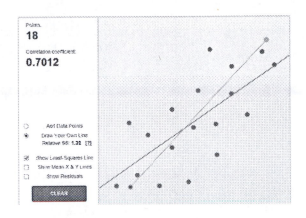

2.113 The plot shown is a very simplified (and not very realistic) example. Circles are economists in business; squares are teaching economists. The plot should show positive association when either set is viewed separately and should show a large number of bachelor's degree economists in business and graduate degree economists in academia.

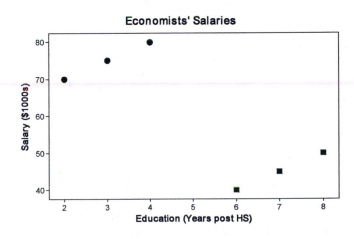

2.115 Sum the rows of the table. $861 + 417 = 1278$ met the requirements and $194 + 557 = 751$ did not meet requirements.

2.117 Divide the cell count by the total for the table. $417/2029 = 0.2055$.

2.119 Divide the cell count by the total number of 11- to 13-year-olds; express that in percent. $417/974 = 0.4281$ (which rounds to 43%).

2.121 (a) Because they want to see the effect of driver's education courses, that is the explanatory variable. The number of accidents is the response. **(b)** Driver's Ed would be the column (x) variable, and number of accidents would be the row (y) variable. A possible table is shown at right. **(c)** There are six cells (two columns by three rows). For example, the first row, first column entry could be the number who took driver's education and had 0 accidents.

	Driver's Ed	
	Yes	No
0 accidents		
1 accident		
2+ accidents		

2.123 (a) Age is the explanatory variable. Rejected is the response. With the dentistry available at that time, it's reasonable to think that as a person got older, he would have lost more teeth.
(b)

	Under 20	20 to 25	25 to 30	30 to 35	35 to 40	Over 40
Yes	0.0002	0.0019	0.0033	0.0053	0.0086	0.0114
No	0.1761	0.2333	0.1663	0.1316	0.1423	0.1196

(c)

Marginal Distribution
Of Rejected?

Yes	No
0.03081	0.96919

Marginal Distribution of Age

Under 20	20 to 25	25 to 30	30 to 35	35 to 40	Over 40
0.1763	0.2352	0.1696	0.1369	0.1509	0.1310

(d) The conditional distribution of Rejected given Age, because we have said Age is the explanatory variable. **(e)** In the table, note that all columns sum to 1. We can clearly see the proportion of rejected recruits increasing with increasing age.

	Under 20	20 to 25	25 to 30	30 to 35	35 to 40	Over 40
Yes	0.0012	0.0082	0.0196	0.0389	0.0572	0.0868
No	0.9988	0.9918	0.9804	0.9611	0.9428	0.9132

2.125 Answers will vary. For example, if a student has a GPA less than 2.0, they are much more likely to enroll for 11 or fewer credits (68.5%). Students with GPAs above 3.0 are most likely to enroll for 15 or more credits (66.6%).

2.127 (a) There are $151 + 148 = 299$ "high exercisers," of which $151/299 = 50.5\%$ get enough sleep and 49.5% (the rest) do not. **(b)** There are $115 + 242 = 357$ "low exercisers," of which $115/357 = 32.2\%$ get enough sleep and 67.8% (the rest) do not. **(c)** Those who exercise more than the median are more likely to get enough sleep.
    Note: *This question is asking for the conditional distribution of sleep within each exercise group. Exercise 2.128 asks for the conditional distribution of exercise within each sleep group.*

2.129 $63/2100 = 3.0\%$ of Hospital A's patients died, compared with $16/800 = 2.0\%$ at B.

2.131 Two examples are shown on the right. In general, choose *a* to be any number from 0 to 200, and then all the other entries can be determined.

50	150
150	50

175	25
25	175

    Note: *This is why we say that such a table has "one degree of freedom": We can make one (nearly) arbitrary choice for the first number, and then have no more decisions to make.*

2.133 Answers will vary. For example, causation might be a negative association between the setting on a stove and the time required to boil a pot of water (higher setting, less time). Common response might be a positive association between SAT score and grade point average. Both of these will have a positive relationship with a person's IQ. An example of confounding might be a negative association between hours of TV watching and grade point average. Once again, people who are naturally smart could finish required work faster and have more time for TV; those who aren't as smart could become frustrated and watch TV instead of doing homework.

2.135 This is a case of confounding: The association between dietary iron and anemia is difficult to detect because malaria and helminths also affect iron levels in the body.

2.137 Responses will vary. For example, students who choose the online course might have more self-motivation or better computer skills. A diagram is shown on the right; the generic "Student characteristics" might be replaced with something more specific.

2.139 No; self-confidence and improving fitness could be common responses to some other personality trait, or high self-confidence could make a person more likely to join the exercise program.

2.141 Patients suffering from more serious illnesses are more likely to go to larger hospitals (which may have more or better facilities) for treatment. They are also likely to require more time to recuperate afterward.

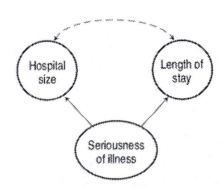

2.143. In this case, there may be a causative effect, but in the direction opposite to the one suggested: people who are overweight are more likely to be on diets and so choose artificial sweeteners over sugar. (Also, heavier people are at a higher risk to develop diabetes; if they do, they are likely to switch to artificial sweeteners.)

2.145. This is an observational study—students choose their "treatment" (to take or not take the refresher sessions).

2.147 (a) The tables are shown below.

Female Titanic Passengers
Class

	1	2	3	Total
Survived	139	94	106	339
Died	5	12	110	127
Total	144	106	216	466

Male Titanic Passengers
Class

	1	2	3	Total
Survived	61	25	75	161
Died	118	146	418	682
Total	179	171	493	843

(b) If we look at the conditional distribution of survival given class for females, 139/144 = 96.53% of first-class females survived, 94/106 = 88.68% survival among second-class females, and 106/216 = 49.07% survival among third-class females. Survival depended on class. **(c)** For males, 61/179 = 34.08% survival among first-class, 25/171 = 14.62% survival among second-class, and 75/493 = 15.21% survival among third-class. Once again, survival depended on class. **(d)** Females overall had much higher survival rates than males.

2.149 (a) Answers will vary, but this is a negative relationship, mostly due to two outliers (Northwest Territories and Nunavut). If those two provinces are deleted, there will be almost no relationship. **(b)** $r = -0.839$. This would not be a good numerical summary for this relationship.

2.151 (b) In the scatterplot at right, we can see that the three territories consistently have a smaller proportion of their populations over 65 than the provinces. The two areas with the largest percent of the population under 15 are Nunavut and the Northwest Territories.

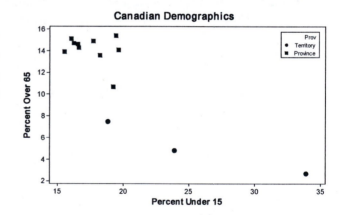

2.153 (a) The relationship is weakly increasing and linear. There seems to be almost two sets of data: five countries with high production and the rest. One country with approximately 225 Dwelling Permit Index might be influential. **(b)** The equation is

Production = 110.96 + 0.0732 DwellPermit.
(c) For 160 dwelling permits, we predict 110.96 + 0.0732*160 = 122.672 for the

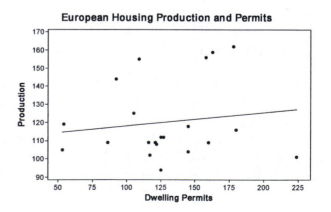

Production Index. **(d)** $e = 109 - 122.672 = -13.672$. **(e)** $r^2 = 2.0\%$. Both indicate very weak relationships, but this is weaker.

2.155 The stacked bar chart clearly shows that offering the RDC service depends on size of the bank. Larger banks are much more likely to offer the service than smaller ones.

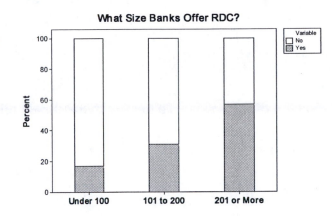

2.157 (a)

Field	Canada	France	Germany	Italy	Japan	UK	US	Total
SsBL	64	153	66	125	250	152	878	1688
SME	35	111	66	80	136	128	355	911
AH	27	74	33	42	123	105	397	801
Ed	20	45	18	16	39	14	167	319
Other	30	289	35	58	97	76	272	857
Total	176	672	218	321	645	475	2069	4576

(b) The marginal distribution of Country is

Canada	France	Germany	Italy	Japan	UK	US
0.0385	0.1469	0.0476	0.0701	0.141	0.1038	0.4521

(c) The marginal distribution for Field of Study is

SsBL	SME	AH	Ed	Other
0.3689	0.1991	0.1750	0.0697	0.1873

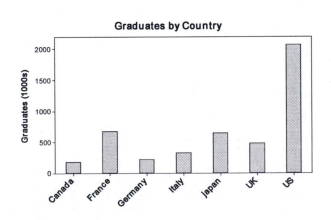

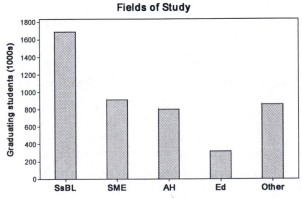

2.159 A school that accepts weaker students but graduates a higher-than-expected number of them would have a positive residual, whereas a school with a stronger incoming class but a lower-than-expected graduation rate would have a negative residual. It seems reasonable to measure school quality by how much benefit students receive from attending the school.

2.161 Answers will vary.

2.163 (a) The residuals are positive at the beginning and end, and negative in the middle. **(b)** The behavior of the residuals agrees with the curved relationship seen in Figure 2.34.

2.165 (a) The regression equation for predicting salary from year is $\widehat{Salary} = 41.253 + 3.9331\,Year$; for year 25, the predicted salary is $41.253 + 3.9331*25 = 139.58$ thousand dollars, or about \$139,600. **(b)** The log-salary regression equation is $\widehat{\ln Salary} = 3.8675 + 0.04832\,Year$. At year 25, we predict $3.8675 + 0.04832*25 = 5.0755$, so the predicted salary is $e^{5.0755} = 160.052$, or about \$160,050. **(c)** Although both predictions involve extrapolation, the second is more reliable because it is based on a linear fit to a linear relationship. **(d)** Interpreting relationships without a plot is risky. **(e)** Student summaries will vary, but should include comments about the importance of looking at plots and the risks of extrapolation.

2.167 (a) The regression equation is $\widehat{2013Salary} = 6523 + 0.9729 \times 2012\,Salary$. **(b)** The residuals appear rather random, but we note the largest positive residuals are on either end of the scatterplot. The largest negative residual is for the next-to-highest 2012–13 salaried person.

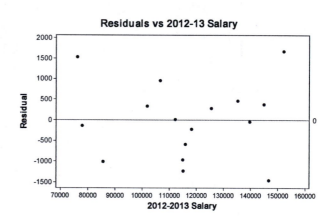

2.169 *Number of firefighters* and *amount of damage* are common responses to the seriousness of the fire.

2.171 (b) The regression line $\widehat{PctCollEd} = 4.033 + 0.906\,FruitVeg5$ generally describes the relationship. There is one outlier at the upper right of the scatterplot (Washington, DC). **(d)** While the scatterplot and regression support a positive association between college degrees and eating fruits and vegetables, association is not causation.

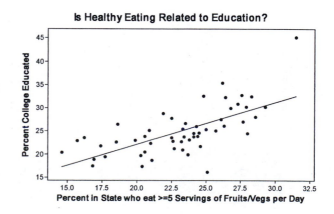

2.173 On the right is a scatterplot of MOR against MOE, showing a moderate linear positive association. The regression equation is $\widehat{MOR} = 2653 + 0.004742MOE$; this regression explains $r^2 = 0.6217 = 62\%$ of the variation in MOR. So, we can use MOE to get fairly good (though not perfect) predictions of MOR.

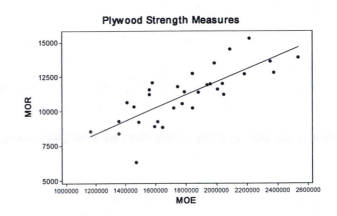

2.175 (a) At right. **(b)** 490/800 = 61.25% of male applicants are admitted, while only 400/700 = 57.14% of females are admitted. **(c)** 400/600 = 66.67% of male business school applicants are admitted; for females, this rate is the same: 200/300 = 66.67%. In the law school, 90/200 = 45% of males are admitted, compared to 200/400 = 50% of females. **(d)** A majority (6/7) of male applicants apply to the business school, which admits (400+200)/(600+300) = 600/900 = 66.67% of all applicants. Meanwhile, a majority (4/7) of women apply to the law school, which admits only (90+200)/(200+400) = 290/600 = 48.33% of its applicants.

	Admit	Deny
Male	490	310
Female	400	300

2.177 If we ignore the "year" classification, we see that Department A teaches 32 small classes out of 52, or about 61.54%, whereas Department B teaches 42 small classes out of 106, or about 39.62%. (These agree with the dean's numbers.)

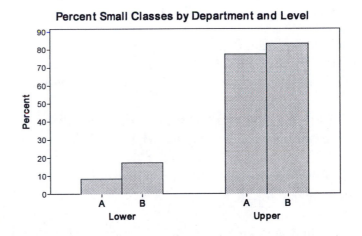

For the report to the dean, students may analyze the numbers in a variety of ways, some valid and some not. The key observations are: (i) When considering only first- and second-year classes, A has fewer small classes (1/12 = 8.33%) than B (12/70 = 17.14%). Likewise, when considering only upper-level classes, A has 31/40 = 77.5% and B has 30/36 = 83.33% small classes. The graph on the right illustrates this. These numbers are given in the back of the text, so most students should include this in their analysis! (ii) 40/52 = 77.78% of A's classes are upper-level courses, compared to 36/106 = 33.96% of B's classes.

Chapter 3 Solutions

3.1 There could be many reasons. She may have already contracted the flu and didn't know it, for example.

3.3 A statement simply that (s)he didn't use performance-enhancing drugs is meaningless. (Consider Lance Armstrong and other athletes.)

3.5. For example, who owns the Web site? Do they have data to back up this statement, and, if so, what was the source of that data?

3.7 Yes to both. Here are two examples. Pew Research regularly makes available their data (about 6 months after the survey), for academic purposes; surveys are observational. The "warnings and contraindications" packed with medicines are based on experiments.

3.9 This is an observational study. All the soap dispensers were run until their batteries died; there were no treatments.

3.11 This is an experiment. Explanatory variable: apple form (juice or whole fruit); response variable: how full the subject felt.

3.13 In this instance, random samples of tuna cans were found and measured (presumably from California, but perhaps other places as well). The sample would have been large enough to convince the producers that they were underfilling the cans relative to the labeled contents.

3.15 (a) Anecdotal data. The three upset students do not necessarily represent the larger student body. **(b)** This is a sample survey, but it is likely biased (who will bother to go to the Web site and respond?). **(c)** This is still a survey, but random. **(d)** Answers will vary.

3.17 For the milk experiment of Exercise 3.14, there were at least two treatments: no extra milk, and extra milk. There may have been additional treatments in terms of how much extra milk, but we were not told. For the echinacea experiment in Exercise 3.16, the treatments were (1) no pills, (2) pills with no echinacea, (3) pills that had echinacea but were not labeled as such, and (4) pills that had echinacea and were labeled as containing it.

3.19 The treatments were the four coaching types actively assigned to 204 people. The factor is type of coaching with the following levels: increase fruit and vegetable intake and physical activity; decrease fat and sedentary leisure; decrease fat and increase physical activity; and increase fruit and vegetable intake and decrease sedentary leisure. The response is the measure of diet and activity improvement after 3 weeks. This experiment had a very high completion rate, so our interpretation of the results should not be subject to any problems.

3.21

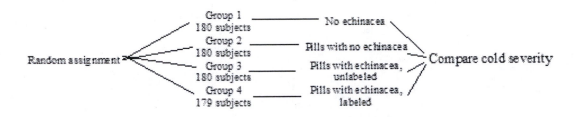

3.23 Using a computer, answers will vary. Each treatment group should end up with 25 subjects.

3.25 (a) Experimental units were the 30 students. They are human, so we can use "subjects." **(b)** We have only one "treatment," so the experiment is not comparative. One possibility is to randomly assign half of the students to the online homework system and the other half to "standard" homework. **(c)** One possibility is grade on an exam over the material from that month; compare those with the online homework to those with "standard" homework.

3.27 (a) Experimental units (subjects): people who go to the Web site. Treatments: description of comfort or showing discounted price. Response variable: shoe sales. **(b)** Comparative, because we have two treatments. **(c)** One option to improve: randomly assign morning and afternoon treatments. **(d)** Yes, a placebo (no special description or price) could give a "baseline" sales figure.

3.29 Starting on line 101, using 1-5 as morning and 6-0 as afternoon for comfort description, we have 19223 95034 comfort gets afternoons on days 2, 6, and 8, and mornings the other days. Starting on other lines will result in different randomizations.

3.31 Yes, each customer (who returns) will get both treatments. However, if a customer accesses the site more than twice, it is no longer matched pairs.

3.33 (a) Shopping patterns may differ on Friday and Saturday. **(b)** Responses may vary in different states. **(c)** A control is needed for comparison.

3.35 For example, new employees should be randomly assigned to either the current program or the new one. One possible outcome would be whether the new employee is still with the company 6 months later.

3.37 (a) The factors are calcium dose and vitamin D dose. There are nine treatments (each calcium/vitamin D combination). **(b)** Assign 20 students to each group, with 10 of each gender. The complete diagram (including the blocking step) would have a total of 18 branches. On the next page is a portion of that diagram, showing only three of the nine branches for each gender. **(c)** Randomization results will vary. **(d)** Yes, there is a placebo. The group that gets 0 mg of both calcium and vitamin D serves as the placebo group.

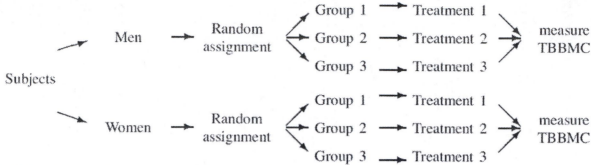

3.39 As described, there are two factors: ZIP code (three levels: none, five-digit, nine-digit) and the day on which the letter is mailed (three levels: Monday, Thursday, or Saturday) for a total of nine treatments. To control lurking variables, aside from mailing all letters to the same address, all letters should be the same size and either printed in the same handwriting or typed. The design should also specify how many letters will be in each treatment group. Also, the letters should be sent randomly over many weeks.

3.41 **(a)** The results of the first random selection are shown below.

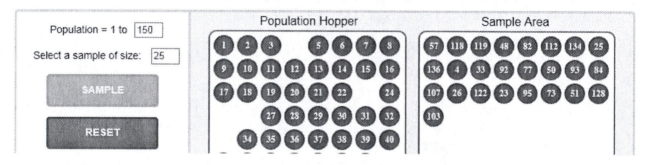

(b) Click **Sample** again for the second 25. They will begin immediately after the end of the first sample (103 in the results shown above). **(c)** Continue until 125 have been assigned to a group,

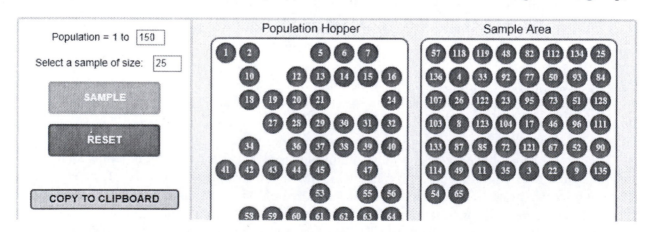

and check that there are 25 remaining in the "Population Hopper."

3.43 Design (a) is an experiment. Because the treatment is randomly assigned, the effect of other habits would be "diluted" because they would be more-or-less equally split between the two groups. Therefore, any difference in colon health between the two groups could be attributed to the treatment (bee pollen or not).

Design (b) is an observational study. It is flawed because the women observed choose whether or not to take bee pollen; one might reasonably expect that people who choose to take bee pollen have other dietary or health habits that would differ from those who do not take the pollen.

3.45 (a) Randomly assign half the girls to get high-calcium punch; the other half will get low-calcium punch. The response variable is not clearly described in this exercise; the best we can say is "observe how the calcium is processed." **(b)** Randomly select half of the girls to receive high-calcium punch first (and low-calcium punch later), while the other half gets low-calcium punch first (followed by high-calcium punch). For each subject, compute the difference in the response variable for each level. This is a better design because it deals with person-to-person variation; the differences in responses for 40 individuals give more precise results than the difference in the average responses for two groups of 20 subjects. **(c)** The first five subjects are 35, 39, 16, 04, and 26. In the completely randomized design, the first group receives high-calcium punch all summer; in the matched pairs design, they receive high-calcium punch for the first part of the summer, and then low-calcium punch in the second half.

3.47 Answers will vary. For example, the trainees and experienced professionals could evaluate the same water samples.

3.49 The population is forest owners from this region. The 348 returned questionnaires are the sample. Response rate: $348/772 = 45\%$. Additionally, we would like to know the sample design (among other things).

   Note: *It would also be reasonable to consider the 772 who received questionnaires as the sample, but we do not get information from all of them.*

3.51 See the solution to the previous exercise; for this problem, we need to choose three items instead of two, but the set up is otherwise the same.

3.53 (a) The population is fans who go to the college's football games (or, especially season ticket holders—those form the sampling frame). **(b)** The sample is the 150 who were sent the questionnaire (or the 98 who responded, from whom they got information). **(c)** The response rate was 65% (98/150). **(d)** The nonresponse rate was 35% (52/150). **(e)** The response rate was pretty good. One possibility to increase the rate might be to offer free food at the next home game, or hold a drawing from respondents for a free season ticket for next year.

3.55 (a) Answers will vary depending on use of software. **(b)** Software is usually more efficient than Table B (you'll have to skip over many possible two-digit labels to find ones between 01 and 18).

3.57 (a) This statement confuses the ideas of population and sample. (If the entire population is found in our sample, we have a *census* rather than a sample.) **(b)** "Dihydrogen monoxide" is H_2O

(water). Any concern about the dangers posed by water most likely means that the respondent did not know what dihydrogen monoxide was and was too embarrassed to admit it. (Conceivably, the respondent could have known the question was about water and had concerns arising from a bad experience of flood damage or near-drowning. But misunderstanding seems to be more likely.) **(c)** Honest answers to such questions are difficult to obtain even in an anonymous survey; in a public setting like this, it would be surprising if there were any raised hands (even though there are likely to be at least a few cheaters in the room).

3.59 The population is (all) local businesses. The sample is the 72 businesses that return the questionnaire, *or* the 160 businesses selected. The nonresponse rate is 55% = 88/160.

   Note: *The definition of "sample" makes it somewhat unclear whether the sample includes all the businesses selected or only those that responded. This author's inclination is toward the latter (the smaller group), which is consistent with the idea that the sample is "a part of the population that we actually examine."*

3.61 Note that the numbers add to 100% down the columns; that is, 39% is the percent of Fox viewers who are Republicans, *not* the percent of Republicans who watch Fox. Students might display the data using a stacked bar graph like the one below, or side-by-side bars. (They could also make four pie charts, but comparing slices across pie charts is difficult.) The most obvious observation is that the party identification of Fox's audience is noticeably different from the other three sources.

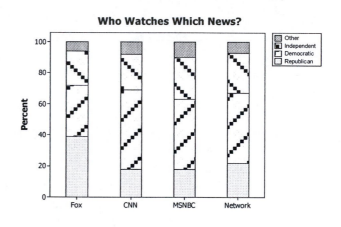

3.63 Labeled in alphabetical order, using line 126, we select: 31 (Village Manor), 08 (Burberry), 19 (Franklin Park), 03 (Beau Jardin), and 25 (Pemberley Courts).

3.65 Population = 1 to **200**, Select a sample of size **20**, then click **Reset**, and **Sample**. One such sample is shown, but each will be different.

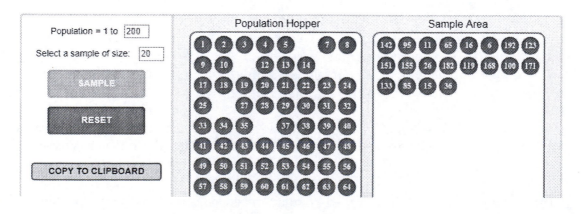

3.67 One could use the labels already assigned to the blocks, but that would mean skipping a lot of four-digit combinations that do not correspond to any block. An alternative would be to drop the second digit and use labels 100–105, 200–211, and 300–325. But by far the simplest approach is to assign labels 01–44 (in numerical order by the four-digit numbers already assigned), enter the table at line 125, and select: 21 (block 3002), 37 (block 3018), 18 (block 2011), 44, 23, 19, 10, 33, and 31.

3.69 Answers will vary. Beginning on line 110, from Group 1 (labeled 1 through 6), select 3 and 4. Continuing from there, from Group 2 (labeled 01 through 12), select 08 and 05. Continuing from there, from Group 3 (labeled 01 through 26), select 13, 09, and 04.

3.71 The sample is random because the starting point is randomly selected (so every individual has an equal chance to be selected before the process begins). Once the random starting point has been selected, the rest of the sample is determined. There is no possibility of selecting (in the previous exercise), students 04 and 05, which could happen in a simple random sample.

3.73 Assign labels 01–46 for the Climax 1 group, 01–62 for the Climax 2 group, and so on. Then, beginning at line 130, select Climax 1: 05, 16, 17, 40, and 20. Climax 2: 19, 45, 05, 32, 19, and 41. Climax 3: 04, 19, and 25. Secondary: 29, 20, 43, and 16.

3.75 Each student has a 12.5% chance: 4 out of 32 over-21 students, and 2 of 16 under-21 students. This is not an SRS because not every group of five students can be chosen; the only possible samples are those with three older and two younger students.

3.77 (a) This design would omit households without telephones or with unlisted numbers. Such households would likely be made up of poor individuals (who cannot afford a phone), those who choose not to have phones (perhaps because they use a cell phone exclusively), and those who do not wish to have their phone numbers published. **(b)** Those with unlisted numbers would be included in the sampling frame when a random-digit dialer is used.

3.79 The female and male students who responded are the samples. The populations are all college undergraduates (males and females) who could be judged to be similar to the respondents. This report is incomplete; a better one would give numbers who responded, as well as the actual response rate.

3.81 The larger sample would have less sampling variability. (That is, the results would have a higher probability of being closer to the "truth.")

3.83 The rand command in Excel will enter one random number between 0 and 1 with **=rand()**. You can select that cell and drag it down until you have 100 such random numbers. Repeat for the second column. Enter **=(a1+b1)/2** in the first cell of the third column. Then, drag that formula down to find the means for all 100 rows. However, Excel doesn't make histograms or stemplots (at least, not without an add-in). Using Minitab to do the same procedures, we can use **Calc > Random Data > Uniform** and complete the dialog box as shown to generate both random samples.   Calc > Row statistics will bring the second dialog box, where you can select the statistic of interest, the input columns, and the output column.

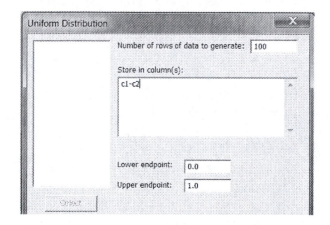

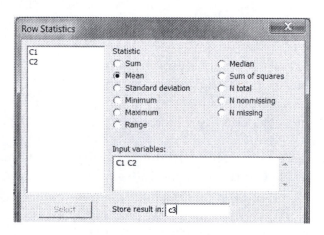

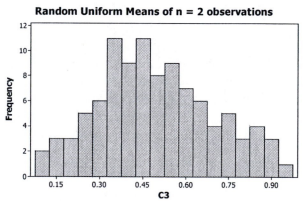

(b) Now you can graph the means. We have shown a histogram. This particular distribution looks right-skewed with variability from about 0.10 to 0.95. The center is about 0.45. **(c)** Means will vary but should be close to 0.5. This distribution had mean 0.4983. **(d)** Standard deviations will vary as well, but they should be close to $1/\sqrt{24} = 0.2041$. This distribution had $s = 0.2004$.

3.85 Now, one will simulate the data into columns 1 through 12 and then compute the row-wise means. This distribution is more symmetric than the one in the solution to Exercise 3.83, but it has small "peaks" at each end. Both small and large sample means occurred more often than would be expected in a truly Normal distribution. We note there is less variability; these sample means range from roughly 0.35 to 0.70. **(b)** The mean of the sample means is 0.5104 (close to 0.5). **(c)** The standard deviation of the sample means is $s = 0.0775$ (theory says this should be 0.0833).

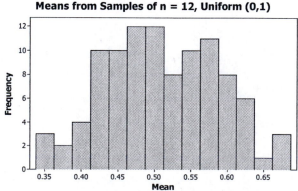

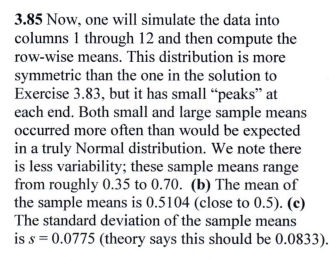

3.87 We have chosen to do 1000 simulations (to try for a "regular" looking histogram). This histogram looks very Normal and ranges essentially from −3 to 3 with a center about 0 (the mean is actually −0.0370) and standard deviation $s = 1.0034$. This appears to be an adequate simulation of a standard Normal distribution.

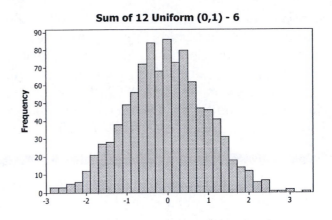

3.89 (a) The simulation of the mean of 12 random observations should have a smaller variability than the mean of two observations. **(b)** The simulations in the solutions to Exercises 3.83 and 3.85 do confirm this (see details above).

3.91 (a) Population: all students at 4-year colleges in the United States Sample: 17,096 students. **(b)** Population: all restaurant workers. Sample: 100 workers. **(c)** Population: all 584 longleaf pine trees. Sample: 40 trees.

3.93 Results will vary. For example, four repetitions with $p = 0.6$ resulted in proportions of 0.56, 0.68, 0.40, and 0.56 heads. Students should continue until they have 50 samples using $p = 0.6$ then make a histogram of the sample proportions. The histogram should be centered at (roughly) 0.6 with standard deviation 0.98. **(b)** This repeats the process using $p = 0.2$. The histogram should be centered at about 0.2 with standard deviation about 0.08.

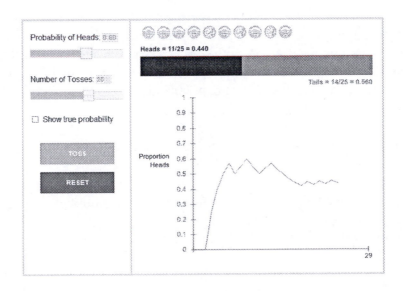

3.95 Student results will vary greatly, and 10 values of $\bar{x}$ will give little indication of the appearance of the sampling distribution. In fact, the sampling distribution of $\bar{x}$ is approximately Normal with a mean of 50.5 and a standard deviation of about 8.92; this approximating Normal distribution is shown on the right (above). Therefore, nearly every sample of size 10 would yield a mean between 23 and 78.

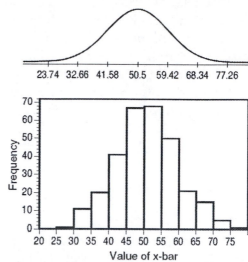

The shape of the sampling distribution becomes more apparent if the results of many students are pooled. Here, we give an example based on 300 sample means, which might arise from pooling all the results in a class of 30.

Note: *Because the values in these samples are not independent (there can be no repeats), a stronger version of the central limit theorem is needed to determine that the sampling distribution is approximately Normal. Confirming the standard deviation given above is a reasonably difficult exercise even for a mathematics major.*

Many of the questions in Section 3.5 (Ethics), Exercises 3.96–3.117, are matters of opinion and may be better used for class discussion rather than as assigned homework. A few comments are included here.

3.97 (a) A nonscientist might raise different viewpoints and concerns from those considered by scientists. **(b)** Answers will vary.

3.107 The articles are "Facebook and academic performance: Reconciling a media sensation with data" (Josh Pasek, eian more, Eszter Hargittai), a critique of the first article called "A response to reconciling a media sensation with data" (Aryn C. Karpinski), and a response to the critique ("Some clarifications on the Facebook-GPA study and Karpinski's response") by the original authors. In case these articles are not available at the address given in the text, they might be found elsewhere with a Web search.

3.109 To control for changes in the mass spectrometer over time, we should alternate between control and cancer samples.

3.111 They cannot be anonymous because the interviews are conducted in person in the subject's home. They are certainly kept confidential.

    Note: *For more information about this survey, see the GSS Web site:* www.norc.org/GSS+Website

3.115 (a) Those being surveyed should be told the kind of questions they will be asked and the approximate amount of time required. **(b)** Giving the name and address of the organization may give the respondents a sense that they have an avenue to complain should they feel offended or mistreated by the pollster. **(c)** At the time that the questions are being asked, knowing who is paying for a poll may introduce bias, perhaps due to nonresponse (not wanting to give what might be considered a "wrong" answer). When information about a poll is made public, though, the poll's sponsor should be announced.

3.121 (a) You need information about a random selection of his games, not just the ones he chooses to talk about. **(b)** These students may have chosen to sit in the front; all students should be randomly assigned to their seats.

3.123 This is an experiment because each subject is (randomly, we assume) assigned to a treatment. The explanatory variable is the price history seen by the subject (steady prices or fluctuating prices), and the response variable is the price the subject expects to pay.

3.125 Answers will vary. Students are asked to give their own examples.

3.127 The two factors are gear (three levels) and steepness of the course (number of levels not specified). Assuming there are at least three steepness levels—which seems like the smallest reasonable choice—that means at least nine treatments. Randomization should be used to determine the order in which the treatments are applied. Note that we must allow ample recovery time between trials, and it would be best to have the rider try each treatment several times.

3.129. (a) One possible population: all full-time undergraduate students in the fall term on a list provided by the registrar. **(b)** A stratified sample with 125 students from each class rank is one possibility. **(c)** Mailed (or emailed) questionnaires might have high nonresponse rates. Telephone interviews exclude those without phones and may mean repeated calling for those who do not answer. Face-to-face interviews might be more costly than your funding will allow. There might also be some response bias: Some students might be hesitant about criticizing the faculty (while others might be far too eager to do so).

3.131 Use a block design: Separate men and women, and randomly allocate each gender among the six treatments.

The remaining exercises relate to the material of Section 3.5 (Ethics). Answers are given for the first two; the rest call for student opinions, or information specific to the student's institution.

3.133 The latter method (CASI) will show a higher percentage of drug use because respondents will generally be more comfortable (and more assured of anonymity) about revealing illegal or embarrassing behavior to a computer than to a person, so they will be more likely to be honest.

Chapter 4 Solutions

4.1 Ten of the first 20 digits on line 121 correspond to "heads," so the proportion of heads is 10/20 = 0.5. While this sample obtained the intuitive half heads and half tails, with such a small sample, random variation can produce results different from the expected value (0.5). For example, the first 20 digits of line 122 give us only 8 heads, so the proportion of heads is 8/20 = 0.4.

```
71487  09984  29077  14863
TTHHT  HTTHH  HTHTT  THHHT
```

4.3 (a) We can discuss the probability (chance) the temperature would be between 30 and 35 degrees Fahrenheit, for example. **(b)** Depending on your school, student identification numbers are probably not random. For example, at the author's university, all student IDs begin with 900, which means that the first three digits are all the same and not random. **(c)** The probability of an ace in a single draw is 1/52 if the deck is well-shuffled.

4.5 Answers will vary depending on your set of 25 rolls.

4.7 If you hear music (or talking) one time, you will almost certainly hear the same thing for several more checks after that. (For example, if you tune in at the beginning of a 5-minute song and check back every 5 seconds, you'll hear that same song over 30 times.)

4.9 Answers will vary. This particular set of rolls had 21/40 rolls that included at least one six. Continue until you are convinced of the result.

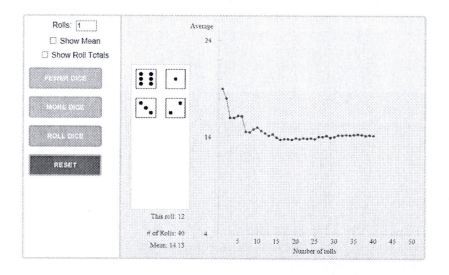

4.11 One possibility: from 0 to 168 hours (the largest number should be big enough to include all possible responses, although we hope that no student texts all day, every day for an entire week!). In addition, some students might respond with fractional answers (e.g., 3.5 hours).

4.13 A favorite color being white, black, silver, gray or red means is it not blue, brown, or other. This probability is 1 − (0.09 + 0.05 + 0.06) = 1 − 0.20 = 0.80. Adding three probabilities and subtracting that result from 1 is slightly easier than adding the five probabilities of interest.

4.15 These are disjoint events, so $P(5 \text{ or } 3 \text{ or less}) = P(5) + P(3 \text{ or less}) = 0.079 + (0.301 + 0.176 + 0.125) = 0.079 + 0.602 = 0.681$.

4.17 Heads and tails are equally likely on each toss, so the sample space of outcomes is {HH, HT, TH, TT}. Getting a head and then a tail is one of these, so the probability is $1/4 = 0.25$.

4.19 (a) $S = \{\text{Yes, No}\}$. **(b)** $S = \{0, 1, 2, ..., x\}$, where x is a reasonable upper limit. **(c)** $S = \{18, 19, 20, ...\}$ (presuming the friends of college students are most likely other college students). There is some leeway here with the lower and upper ends of the ages. **(d)** Answers will vary by institution. Some have many (for example, Purdue University has almost 200 possible majors), other have few (for example, a culinary school presumably has just one major).

4.21 (a) Not equally likely (Check the Web site. As of November 10, 2013, Azarenka's win-loss record is 43-9. See http://sports.espn.go.com/tennis/player/_/id/421/victoria-azarenka). **(b)** Equally likely. The chance of a king is 4/52 and the chance of a two is 4/52. **(c)** This could depend on the intersection; for example, is the turn onto a one-way street? **(d)** Not equally likely. The home team "usually" wins—this is home court advantage. (For example, see http://espn.go.com/mens-college-basketball/story/_/id/8848242/nation-best-homecourt-advantages-college-basketball for a discussion about home court advantage.)

4.23 (a) The probability that both of the two disjoint events occur is 0. (Multiplication of probabilities is appropriate for *independent* events, not *disjoint* events.) **(b)** Probabilities must be no more than 1; $P(A \text{ and } B)$ will be no more than 0.5. (We cannot determine this probability exactly from the given information.) **(c)** $P(\text{not } A) = 1 - P(A) = 1 - 0.35 = 0.65$.

4.25 There are six possible outcomes: $S = \{\text{link1, link2, link3, link4, link5, leave}\}$.

4.27 (a) $P(\text{Wanted or Gangnam Style}) = 0.086 + 0.086 = 0.172$. **(b)** $P(\text{not (Wanted or Gangnam Style)}) = 1 - 0.172 = 0.828$.

4.29 (a) $P(AB) = 1 - (0.42 + 0.11 + 0.44) = 1 - 0.97 = 0.03$. **(b)** $P(\text{can donate}) = P(O \text{ or } B) = P(O) + P(B) = 0.44 + 0.11 = 0.55$.

4.31 (a) No. These student categories are disjoint and the probabilities sum to more than 1. **(b)** This is legitimate (in terms of the probability rules), but the deck would be a non-standard one. **(c)** This is legitimate, but represents a "loaded" die (i.e., not a fair one).

4.33 (a) $P(\text{some education beyond high school, but no degree}) = 1 - (0.12 + 0.31 + 0.29) = 0.28$. **(b)** $P(\text{at least high school}) = 1 - P(\text{didn't finish high school}) = 1 - 0.12 = 0.88$.

4.35 For example, the probability for A-positive blood is $(0.42)(0.84) = 0.3528$ and for A-negative blood is $(0.42)(0.16) = 0.0672$.

Blood type	A+	A−	B+	B−	AB+	AB−	O+	O−
Probability	0.3528	0.0672	0.0924	0.0176	0.0252	0.0048	0.3696	0.0704

4.37 (a) There are six arrangements of the digits 4, 9, and 1 (491, 419, 941, 914, 149, 194), so that $P(\text{win}) = 6/1000 = 0.006$. **(b)** The only winning arrangement is 222, so $P(\text{win}) = 1/1000 = 0.001$.

4.39 $P(\text{none are O-negative}) = P(\text{all are not O-negative}) = (1 - 0.07)^{10} = 0.4840$, so $P(\text{at least one is O-negative}) = 1 - P(\text{none are O-negative}) = 1 - 0.4840 = 0.5160$.

4.41 Note that $A = (A \text{ and } B)$ or $(A \text{ and } B^c)$, and the events $(A \text{ and } B)$ and $(A \text{ and } B^c)$ are disjoint, so Rule 3 says that $P(A) = P((A \text{ and } B) \text{ or } (A \text{ and } B^c)) = P(A \text{ and } B) + P(A \text{ and } B^c)$. If $P(A \text{ and } B) = P(A)P(B)$, then we have $P(A \text{ and } B^c) = P(A) - P(A)P(B) = P(A)(1 - P(B))$, which equals $P(A)P(B^c)$ by the complement rule.

4.43. (a) Nancy and David's children can have alleles BB, BO, or OO, so they can have blood type B or O. (The table on the right shows the possibilities.) **(b)** Either note that the four combinations in the table are equally likely or compute $P(\text{type O}) = P(\text{O from Nancy and O from David}) = 0.5^2 = 0.25$ and $P(\text{type B}) = 1 - P(\text{type O}) = 0.75$.

	B	O
B	BB	BO
O	BO	OO

4.45. (a) Any child of Jasmine and Joshua has an equal (1/4) chance of having blood type AB, A, B, or O (see the allele combinations in the table). Therefore, $P(\text{type O}) = 0.25$. **(b)** $P(\text{all three have type O}) = 0.25^3 = 0.015625 = 1/64$. $P(\text{first has type O, next two do not}) = 0.25(0.75^2) = 0.140625 = 9/64$.

	A	O
B	AB	BO
O	AO	OO

4.47 Two tosses of a fair coin can result in HH, HT, TH, or TT. Each of these has probability 1/4. Counting the number of heads in the two tosses, we have

x	0	1	2
$P(X=x)$	0.25	0.5	0.25

4.49

x	1	2	3	4	5	6
$P(X=x)$	0.05	0.05	0.13	0.26	0.36	0.15

4.51 (a) $P(X \le 3) = 0.05 + 0.05 + 0.13 = 0.23$. **(b)** $P(X = 4 \text{ or } X = 5) = 0.26 + 0.36 = 0.62$. **(c)** $P(X = 8) = 0$.

4.53 (a) The probabilities for a discrete *random variable* add to 1. **(b)** Continuous random variables can take values from any interval, not just 0 to 1. **(c)** A Normal random variable is continuous. (Also, a distribution is *associated with* a random variable, but "distribution" and "random variable" are not the same things.)

4.55 (a) Based on the information from Exercise 4.50, along with the complement rule, $P(T) = 0.19$ and $P(T^c) = 0.81$. **(b)** Use the multiplication rule for independent events; for example, $P(TTT) = 0.19^3 = 0.0069$, $P(TTT^c) = (0.19^2)(0.81) = 0.0292$, $P(TT^cT^c) = (0.19)(0.81^2) = 0.1247$, and $P(T^cT^cT^c) = 0.81^3 = 0.5314$. **(c)** Add up the probabilities from (b) that correspond to each value of X. Note that we have accounted for rounding in the table on the next page.

Outcome	TTT	TTT^c	TT^cT	T^cTT	TT^cT^c	T^cTT^c	T^cT^cT	$T^cT^cT^c$
Probability	0.0069	0.0292	0.0292	0.0292	0.1247	0.1247	0.1247	0.5314
X	0	1			2			3
Probability	0.0069	0.0876			0.3741			0.5314

4.57 (a) Time is continuous. **(b)** Hits are discrete (you can count them). **(c)** Yearly income is discrete (you can count money).

4.59 (a) The pairs are given below. We must assume that we can distinguish between, for example, "(1,2)" and "(2,1)"; otherwise, the outcomes are not equally likely. **(b)** Each pair has probability 1/36. **(c)** The value of X is given below each pair. For the distribution (given below), we see (for example) that there are four pairs that add to 5, so $P(X = 5) = 4/36$. Histogram below, right. **(d)** $P(7 \text{ or } 11) = 6/36 + 2/36 = 8/36 = 2/9 = 0.22\overline{2}$. **(e)** $P(\text{not } 7) = 1 - 6/36 = 5/6 = 0.8333$.

(1, 1)	(1, 2)	(1, 3)	(1, 4)	(1, 5)	(1, 6)
2	3	4	5	6	7
(2, 1)	(2, 2)	(2, 3)	(2, 4)	(2, 5)	(2, 6)
3	4	5	6	7	8
(3, 1)	(3, 2)	(3, 3)	(3, 4)	(3, 5)	(3, 6)
4	5	6	7	8	9
(4, 1)	(4, 2)	(4, 3)	(4, 4)	(4, 5)	(4, 6)
5	6	7	8	9	10
(5, 1)	(5, 2)	(5, 3)	(5, 4)	(5, 5)	(5, 6)
6	7	8	9	10	11
(6, 1)	(6, 2)	(6, 3)	(6, 4)	(6, 5)	(6, 6)
7	8	9	10	11	12

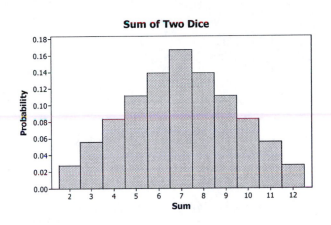

Sum	2	3	4	5	6	7	8	9	10	11	12
Probability	$\frac{1}{36}$	$\frac{2}{36}$	$\frac{3}{36}$	$\frac{4}{36}$	$\frac{5}{36}$	$\frac{6}{36}$	$\frac{5}{36}$	$\frac{4}{36}$	$\frac{3}{36}$	$\frac{2}{36}$	$\frac{1}{36}$

4.61 (a) A histogram is shown at right. **(b)** $X \geq 1$, where X is the number of nonword errors. $P(X \geq 1) = 1 - P(X = 0) = 1 - 0.1 = 0.9$. **(c)** At most, two nonword errors. $P(X \leq 2) = 0.1 + 0.3 + 0.3 = 0.7$. $P(X < 2) = 0.1 + 0.3 = 0.4$.

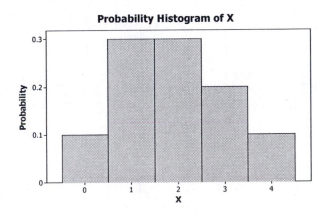

4.63 (a) The height should be $\frac{1}{2}$ since the area under the curve must be 1. The density curve is at the right. **(b)** $P(Y \le 1.6) = 1.6/2 = 0.8$. **(c)** $P(0.5 < Y < 1.7) = 1.2/2 = 0.6$. **(d)** $P(Y \ge 0.95) = 1.05/2 = 0.525$.

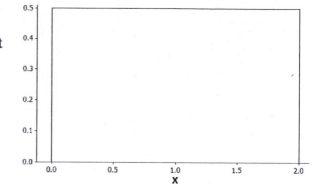

4.65 $P(8 \le X \le 10) = P\left(\dfrac{8-9}{0.0724} \le \dfrac{X-9}{0.0724} \le \dfrac{10-9}{0.0724}\right) = P(-13.8 \le Z \le 13.8)$. This probability is essentially 1; x will almost certainly estimate μ within ± 1 (in fact, it will almost certainly be much closer than this).

4.67 The possible values of X are \$0 and \$5, each with probability 0.5. The mean amount won will be \$0(0.5) + \$5(0.5) = \$2.50.

4.69 Using Rule 1, $\mu_Y = 12 + 7\mu_X = 12 + 7(8) = 68$.

4.71 First, $\mu_X = 0.4(\$0) + 0.6(\$3) = \$1.80$. $\sigma_X^2 = (0-1.80)^2(0.4) + (3-1.80)^2(0.6) = 2.16$. The standard deviation is $\sigma = \sqrt{2.16} = \$1.47$.

4.73 As sample size gets larger, the standard deviation decreases. The mean for 1000 will be much closer to μ than the mean for 2 (or 100) observations. Intuitively, as you sample more of the population, you have better information about the population, so any estimates about the population should get "better," that is, closer to the true value(s).

4.75 From Exercise 4.72, we have $\mu_X = 0.5$.
$\sigma_X^2 = (-1-0.5)^2(0.3) + (0-0.5)^2(0.2) + (1-0.5)^2(0.2) + (2-0.5)^2(0.3) = 2.25(0.3) + 0.25(0.2) + 0.25(0.2) + 2.25(0.3) = 1.45$. $\sigma = \sqrt{1.45} = 1.204$.

4.77 Correlation does not affect means. The means of the variable Z in this exercise are exactly as shown in the solution to Exercise 4.74 above.

4.79 $\mu = 0(0.3) + 1(0.1) + 2(0.1) + 3(0.2) + 4(0.2) + 5(0.1) = 2.2$.

4.81 $\sigma = \sqrt{(0-0.1538)^2(0.8507) + (1-0.1538)^2(0.1448) + (2-0.1538)^2(0.0045)} =$
$\sqrt{0.0201228321 + 0.1036846829 + 0.015338045} = \sqrt{0.013914556} = 0.373$.

   Note: *Any rounding of intermediate terms can result in a significant amount of roundoff error at the end of the calculation.*

4.83 (a) $\sigma = \sqrt{75^2 + 41^2} = \$85.48$. **(b)** This is larger than that calculated in the example. The negative correlation makes the standard deviation of the sum smaller.

4.85 The situation described in this exercise—"people who have high intakes of calcium in their diets are more compliant than those who have low intakes"—implies a positive correlation between calcium intake and compliance. Because of this, the variance of total calcium intake is greater than the variance we would see if there were no correlation (as the calculations in Example 4.39 demonstrate).

4.87 (a) For a single toss, X = number of heads has possible values 0 and 1.
$\mu_X = 0(0.5) + 1(0.5) = 0.5$, and $\sigma_X = \sqrt{(0-0.5)^2(0.5) + (0-0.5)^2(0.5)} = \sqrt{0.25} = 0.5$. **(b)** Tossing four times, we have $Y = X_1 + X_2 + X_3 + X_4$. $\mu_Y = 0.5 + 0.5 + 0.5 + 0.5 = 2$.
$\sigma_Y = \sqrt{0.25 + 0.25 + 0.25 + 0.25} = \sqrt{1} = 1$. **(c)** The use of the distribution is illustrated below. Summing the xp column, we found $\mu = 2$; summing the last column gives $\sigma^2 = 1$. Thus, $\sigma = 1$.

x	$P(X{=}x)$	xp	$(x-\mu)^2 p$
0	0.0625	0	0.25
1	0.25	0.25	0.25
2	0.375	0.75	0
3	0.25	0.75	0.25
4	0.0625	0.25	0.25

Note: *This exercise illustrates an important fact: The result of four tosses of a coin is not the same as multiplying the result of one toss by 4. That idea works for the mean, but not for the variance and standard deviation.*

4.89 (a) Because cards are dealt without replacement, the total points X and Y are not independent. **(b)** The result of one roll will not affect the other; X and Y are independent in this case.

4.91 Although the probability of having to pay for a total loss for one or more of the 10 policies is very small, if this were to happen, it would be financially disastrous. On the other hand, for thousands of policies, the law of large numbers says that the average claim on many policies will be close to the mean, so the insurance company can be assured that the premiums they collect will (almost certainly) cover the claims.

4.93 (a) Add up the given probabilities and subtract from 1; this gives P(man does not die in the next 5 years) = 0.99749. **(b)** The distribution of income (or loss) is given below. Multiplying each possible value by its probability gives the mean intake $\mu = \$623.22$.

Age at death	25	26	27	28	29	Survives
Loss or income	−$99,825	−$99,650	−$99,475	−$99,300	−$99,125	$875
Probability	0.00039	0.00044	0.00051	0.00057	0.00060	0.99749

4.95 $P(2 \text{ or } 4 \text{ or } 5) = P(2) + P(4) + P(5) = 1/6 + 1/6 + 1/6 = 1/2 = 0.5$.

4.97 Let A be the event "next card is an ace" and B be "two of Slim's four cards are aces." Then, $P(A \mid B) = \dfrac{2}{48}$ because (other than the four known cards in Slim's hand) there are 48 cards, of which two are aces.

4.99 This computation uses the addition rule for disjoint events, which is appropriate for this setting because B (full-time students) is made up of four disjoint groups (those in each of the four age groups).

 4.101

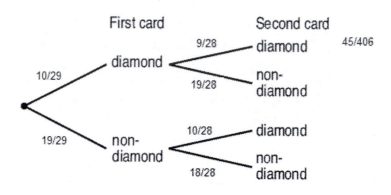

4.103 (a) Because the events are disjoint, $P(A \cap B \cap C) = 0.4 + 0.3 + 0.1 = 0.8$. **(b)** See the diagram. **(c)** $P(A \cap B \cap C)^c = 1 - P(A \cap B \cap C) = 1 - 0.8 = 0.2$.

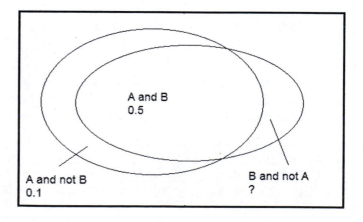

4.105 (a) $P(B \mid A) = 0.5/0.6 = 5/6 = 0.8333$. **(b)** The diagram is given.

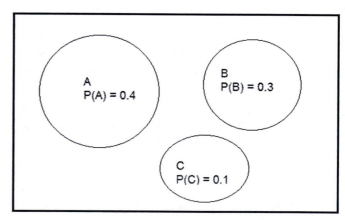

4.107 The letters assigned here are certainly not the only possible combinations. **(a)** Let A = "5 to 10 years old" and A^c = 11 to 13 years old (these are the only two age groups under consideration). Let C = adequate calcium intake and C^c = inadequate calcium intake. **(b)** $P(A)$ = 0.52; $P(A^c)$ = 0.48. $P(C^c \mid A)$ = 0.18; $P(C^c \mid A^c)$ = 0.57. **(c)** The tree diagram is given following. Multiply out the branches to find P(inadequate calcium) = (0.52)(0.18) + (0.48)(0.57) = 0.3672.

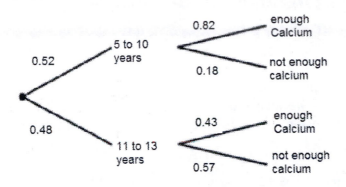

4.109 These events are not independent; the probability of having inadequate calcium intake changes with the age of the child. From the solution to Exercise 4.108, we have P(11 to 13 | inadequate calcium) = 0.7451, which is not the same as P(11 to 13) = 0.48.

4.111 (a) 0.40 – 0.24 = 0.16. **(b)** 0.46 – 0.24 = 0.22. **(c)** 1 – (0.40 + 0.46 – 0.24) = 0.38. **(d)** The answers in (a) and (b) are found by a variation of the addition rule for disjoint events: We note that $P(S) = P(S \text{ and } E) + P(S \text{ and } E^c)$ and $P(E) = P(S \text{ and } E) + P(S^c \text{ and } E)$. In each case, we know the first two probabilities, and we find the third by subtraction. The answer for (c) is found by using the general addition rule to find $P(S \text{ or } E)$, and noting that S^c and $E^c = (S \text{ or } E)^c$.

4.113 For a randomly chosen high school student, let L = "student admits to lying" and M = "student is male," so $P(L)$ = 0.48, $P(M)$ = 0.5, and $P(M \text{ and } L)$ = 0.25. Then, $P(M \text{ or } L)$ = $P(M) + P(L) - P(M \text{ and } L)$ = 0.48 + 0.5 − 0.25 = 0.73.

4.115 Let M = "male" and C = "attends a 4-year institution." (C is not an obvious choice, but it is less confusing than F, which we might mistake for "female.") We have been given $P(C)$ = 0.61, $P(C^c)$ =

	Men	Women
Four-year institution	0.2684	0.3416
Two-year institution	0.1599	0.2301

0.39, $P(M \mid C)$ = 0.44, and $P(M \mid C^c)$ = 0.41. **(a)** To create the table, observe that: $P(M \text{ and } C)$ = $P(M \mid C)P(C)$ = (0.44)(0.61) = 0.2684. Similarly, $P(M \text{ and } C^c)$ = $P(M \mid C^c)P(C^c)$ = (0.41)(0.39) = 0.1599. The other two entries can be found in a similar fashion or by observing that, for example, the two numbers on the first row must sum to $P(C)$ = 0.61.

(b) $P(C \mid M^c) = \dfrac{P(C \text{ and } M^c)}{P(M^c)} = \dfrac{0.3416}{0.3416 + 0.2301} = 0.5975.$

4.117 Let M = "male" and C = "attends a 4-year institution." For this tree diagram, we need to compute $P(M)$ = $0.2684 + 0.1599 = 0.4283$, $P(M^c)$ = $0.3416 + 0.2301 = 0.5717$, as well as $P(C \mid M)$, $P(C \mid M^c)$, $P(C^c \mid M)$, and $P(C^c \mid M^c)$. For example,

$$P(C \mid M) = \frac{P(C \text{ and } M)}{P(M)} = \frac{0.2684}{0.4283} =$$

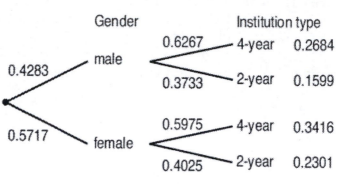

0.6267. All the computations for this diagram are "inconvenient" because they require that we work *backward* from the ending probabilities, instead of working *forward* from the given probabilities (as we did in the previous tree diagram).

4.119 $P(A \mid B) = \dfrac{P(A \text{ and } B)}{P(B)} = \dfrac{0.082}{0.261} = 0.3142$. If A and B were independent, then $P(A \mid B)$ would equal $P(A)$, and also $P(A \text{ and } B)$ would equal the product $P(A)P(B)$.

4.121 (a) "The vehicle is a light truck" = A^c; $P(A^c) = 0.69$. **(b)** "The vehicle is an imported car" = A and B. To find this

	$P(A) = 0.31$	$P(A^c) = \mathbf{0.69}$
$P(B) = 0.22$	$P(A \text{ and } B) = 0.08$	$P(A^c \text{ and } B) = 0.14$
$P(B^c) = \mathbf{0.78}$	$P(A \text{ and } B^c) = 0.23$	$P(A^c \text{ and } B^c) = \mathbf{0.55}$

probability, note that we have been given $P(B^c) = 0.78$ and $P(A^c \text{ and } B^c) = 0.55$. From this we can determine that $78\% - 55\% = 23\%$ of vehicles sold were domestic cars—that is, $P(A \text{ and } B^c) = 0.23$—so $P(A \text{ and } B) = P(A) - P(A \text{ and } B^c) = 0.31 - 0.23 = 0.08$.

   Note: *The table shown here summarizes all that we can determine from the given information* (**bold**).

4.123 We seek $P(\text{at least one offer}) = P(A \text{ or } B \text{ or } C)$; we can find this as $1 - P(\text{no offers}) = 1 - P(A^c \text{ and } B^c \text{ and } C^c)$. We see in the Venn diagram of Exercise 4.122 that this probability is 1.

4.125 For the probability of the New Jersey job, if she gets an offer for the federal job, we want $P(B \mid C) = P(B \text{ and } C)/P(C) = 0.1/0.3 = 1/3 = 0.333$. For the probability of the federal job, if she gets an offer for the New Jersey job, we want $P(C \mid B) = P(B \text{ and } C)/P(B) = 0.1/0.5 = 0.2$.

4.127 (a) $P(M) = 0.3959$ (this is a marginal probability). **(b)** $P(B \mid M) = 0.2641/0.3959 = 0.6671$. **(c)** $P(M)P(B) = (0.3959)(0.6369) = 0.2521$. This is not the same as 0.6671, so the events "Male" and "Bachelor's" are not independent.

4.129 (a) Beth's brother has type *aa*, and he got one allele from each parent. But neither parent is albino, so neither could be type *aa*. **(b)** The table on the right shows the possible combinations, each of which is equally likely, so $P(aa) =$

	A	a
A	AA	Aa
a	Aa	aa

0.25, $P(Aa) = 0.5$, and $P(AA) = 0.25$. **(c)** Beth is either AA or Aa, and $P(AA \mid \text{not } aa) = \dfrac{0.25}{0.75} = \dfrac{1}{3}$, while $P(Aa \mid \text{not } aa) = \dfrac{0.50}{0.75} = \dfrac{2}{3}$.

4.131 Let C be the event "Toni is a carrier," T be "Toni tests positive," and D be "her son has DMD." We have $P(C) = \dfrac{2}{3}$, $P(T \mid C) = 0.7$, and $P(T \mid C^c) = 0.1$. Therefore, $P(T) = P(T \text{ and } C) +$

$P(T \text{ and } C^c) = P(C)\, P(T \mid C) + P(C^c)\, P(T \mid C^c) = \dfrac{2}{3}(0.7) + \dfrac{1}{3}(0.1) = 0.5$, and $P(C \mid T) =$

$\dfrac{P(T \text{ and } C)}{P(C)} = \dfrac{(2/3)(0.7)}{0.5} = \dfrac{14}{15} = 0.9333$.

4.133 When repeated many times, the approximate value of the mean will be close to $\mu = 0.2(-1) + 0.8(2) = 1.4$.

4.135 (a) Because the possible values of X are 1 and 2, the possible values of Y are $4(1^2) - 2 = 2$ with probability 0.4 and $4(2^2) - 2 = 14$ with probability 0.6. **(b)** $\mu_Y = 0.4(2) + 0.6(14) = 9.2$.

$\sigma_Y = \sqrt{(2 - 9.2)^2(0.4) + (14 - 9.2)^2(0.6)} = 5.8788$. **(c)** There are no rules for a quadratic function of a random variable; we must use the definitions.

4.137 (a) $P(A) = 1/36$, $P(B) = 15/36$. **(b)** $P(A) = 1/36$, $P(B) = 15/36$. **(c)** $P(A) = 10/36$, $P(B) = 6/36$. **(d)** $P(A) = 10/36$, $P(B) = 6/36$.

4.139 For each bet, the mean is the winning probability times the winning payout, plus the losing probability times $-\$10$. These are summarized in the table on the right; all mean payoffs equal \$0.

Note: *Alternatively, we can find the mean amount of money we have at the end of the bet. For example, if the point is 4 or 10, we end with either \$30 or \$0, and our expected ending amount is* $\dfrac{1}{3}(\$30) + \dfrac{2}{3}(\$0) = \$10$, *the amount bet.*

Point	Expected Payoff
4 or 10	$\dfrac{1}{3}(+20) + \dfrac{2}{3}(-10) = 0$
5 or 9	$\dfrac{2}{5}(+15) + \dfrac{3}{5}(-10) = 0$
6 or 8	$\dfrac{5}{11}(+12) + \dfrac{6}{11}(-10) = 0$

4.141 (a) This is legitimate because all the probabilities are between 0 and 1 and sum to 1. **(b)** $P(\text{tasters agree}) = 0.03 + 0.07 + 0.25 + 0.20 + 0.06 = 0.61$. **(c)** For Taster 1, add all the entries in the bottom two rows; for Taster 2, add the entries in the two right-most columns. $P(\text{Taster 1 rates higher than 3}) = 0.39$. $P(\text{Taster 2 rates higher than 3}) = 0.39$.

4.143 This is the probability of 19 (independent) losses, followed by a win; by the multiplication rule, this is $(0.994)^{19}(0.006) = 0.005352$.

4.145 With B, M, and D representing the three kinds of degrees, and W meaning the degree recipient was a woman, we have been given:

$$P(B) = 0.69, \qquad P(M) = 0.28, \qquad P(D) = 0.03,$$
$$P(W \mid B) = 0.57, \qquad P(W \mid M) = 0.60, \qquad P(W \mid D) = 0.52.$$

Therefore, we find

$$P(W) = P(W \text{ and } B) + P(W \text{ and } M) + P(W \text{ and } D)$$
$$= P(B)\, P(W \mid B) + P(M)\, P(W \mid M) + P(D)\, P(W \mid D) = 0.5769,$$

So, $P(B \mid W) = \dfrac{P(B \text{ and } W)}{P(W)} = \dfrac{P(B)P(W \mid B)}{P(W)} = \dfrac{0.3933}{0.5769} = 0.6817$

4.147 $P(\text{no point is established}) = \dfrac{12}{36} = \dfrac{1}{3}$.

In Exercise 4.139, the probabilities of winning each odds bet were given as $\dfrac{1}{3}$ for 4 and 10, $\dfrac{2}{5}$ for 5 and 9, and $\dfrac{5}{11}$ for 6 and 8. In the diagram shown on the right, the probabilities are omitted from the individual branches. The probability of winning an odds bet on 4 or 10 (with a net payout of \$20) is $\left(\dfrac{3}{36}\right)\left(\dfrac{1}{3}\right) = \left(\dfrac{1}{36}\right)$. Losing

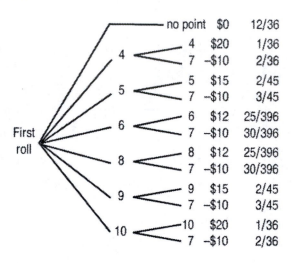

that odds bet costs \$10 and has probability $\left(\dfrac{3}{36}\right)\left(\dfrac{2}{3}\right) = \dfrac{2}{36}$, (or $\dfrac{1}{18}$). Similarly, the probability of

winning an odds bet on 5 or 9 is $\left(\dfrac{4}{36}\right)\left(\dfrac{2}{5}\right) = \dfrac{2}{45}$, and the probability of losing that bet is

$\left(\dfrac{4}{36}\right)\left(\dfrac{3}{5}\right) = \dfrac{3}{45}$, (or $\dfrac{1}{15}$). For an odds bet on 6 or 8, we win \$12 with probability

$\left(\dfrac{5}{36}\right)\left(\dfrac{5}{11}\right) = \dfrac{25}{396}$, and lose \$10 with $\left(\dfrac{5}{36}\right)\left(\dfrac{6}{11}\right) = \dfrac{30}{396} = \dfrac{5}{66}$. To confirm that this game is fair, one

can multiply each payoff by its probability then add up all of those products. More directly, because each individual odds bet is fair (as was shown in the solution to Exercise 4.139), one can argue that taking the odds bet whenever it is available must be fair.

4.149 Let R_1 be Taster 1's rating and R_2 be Taster 2's rating. $P(R_1 = 3) = 0.01 + 0.05 + 0.25 + 0.05 + 0.01 = 0.37$, so:

$$P(R_2 > 3 \mid R_1 = 3) = \frac{P(R_2 > 3 \text{ and } R_1 = 3)}{P(R_1 = 3)} = \frac{0.05 + 0.01}{0.37} = 0.1622.$$

4.151 The event $\{Y < 1/3\}$ is the bottom third of the square, while $\{Y > X\}$ is the upper left triangle of the square. They overlap in a triangle with area 1/18, so:

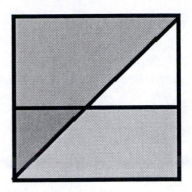

$$P(Y < 1/3 \mid Y > X) = \frac{P(Y < 1/3 \text{ and } Y > X)}{P(Y > X)} = \frac{1/18}{1/2} = \frac{1}{9}.$$

Chapter 5 Solutions

5.1 Population: AppsFire users. Statistic: a median of 108 apps per device. Likely values will vary (there was no indication of variability given).

5.3 $\mu_{\bar{x}} = 420$, $\sigma_{\bar{x}} = \dfrac{\sigma}{\sqrt{n}} = \dfrac{21}{\sqrt{441}} = 1$. When the sample size increases, the mean of the sampling distribution remains the same, but the standard deviation of the sampling distribution decreases.

5.5 The standard deviation of $\bar{x}$ is now $\sigma_{\bar{x}} = \dfrac{\sigma}{\sqrt{n}} = \dfrac{70}{\sqrt{1225}} = 2$. About 95% of the time, $\bar{x}$ is between 181 and 189 (two standard deviations either side of the mean).

5.7 Example 5.6 used the exponential distribution with mean and standard deviation 1.
(a) Each sample size has $\mu_{\bar{x}} = 1$.
For $n = 2$, $\sigma_{\bar{x}} = 1/\sqrt{2} = 0.7071$.
For $n = 10$, $\sigma_{\bar{x}} = 1/\sqrt{10} = 0.3162$.
For $n = 25$, $\sigma_{\bar{x}} = 1/\sqrt{25} = 0.2$.
(b) Shown at right is one realization for $n = 25$.
(c) Answers from the applet will vary, but the mean and standard deviations should be fairly

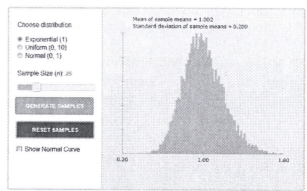

close to the values calculated in part (a). As the sample size increases, the sampling distribution for the sample mean gets less right-skewed and closer to Normal. **(d)** Answers will vary, but generally, $n = 50$ is "large enough" for an exponential random variable to have a sample mean that is approximately Normally distributed.

5.9 (a) The standard deviation for $n = 10$ will be $\sigma_{\bar{x}} = 20/\sqrt{10}$. **(b)** Standard deviation *decreases* with increasing sample size. **(c)** $\mu_{\bar{x}}$ always equals μ and does not depend on the sample size n.

5.11(a) $\mu = 125.5$. **(b)** Answers will vary. **(c)** Answers will vary. **(d)** The center of the histogram represents an average of averages.

5.13. (a) The population in Exercise 5.12 is U.S. smartphone subscribers, while the population in Exercise 5.1 consists of AppsFire users. Presumably, those users with AppsFire will have more apps than those without. **(b)** Excluding those with no apps will increase the median because you are eliminating individuals.

5.15. (a) Larger. For $n = 150$, $2\sigma_{\bar{x}} = 0.202$. **(b)** We need $\sigma_{\bar{x}} \leq 0.17/2 = 0.085$. **(c)** We need $1.24/\sqrt{n} \leq 0.085$. Using algebra, we find $n \geq (1.24/0.085)^2 = 212.817$. The smallest sample size that will fit this criterion is $n = 213$.

5.17 The mean of the sample means will be $\mu_{\bar{x}} = 250$ ml. The standard deviation of the sample means will be $\sigma_{\bar{x}} = 0.5 / \sqrt{4} = 0.25$ ml.

5.19. (a) Graph shown. **(b)** To be more than 1 ml away from the target value means the volume is less than 249 or more than 251. Using symmetry, $P = 2P(X < 249) = 2P(Z < -2) = 2(0.0228) = 0.0456$. **(c)** $P = 2P(X < 249) = 2P(Z < -4)$ which is essentially 0. (Software gives 0.00006.)

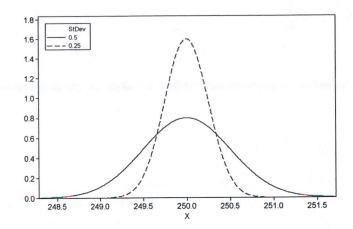

5.21 (a) $\bar{x}$ is not systematically higher than or lower than μ. **(b)** With large samples, $\bar{x}$ is more likely to be close to μ.

5.23 (a) $\mu_{\bar{x}} = 0.3$. $\sigma_{\bar{x}} = 0.8 / \sqrt{100} = 0.08$. **(b)** $z = \dfrac{0.5 - 0.3}{0.08} = 2.5$. From Table A (and technology), $P(\bar{x} > 0.5) = P(Z > 2.5) = 0.0062$. **(c)** $n = 100$ is a large enough sample to be able to use the central limit theorem.

5.25 Let X be Sheila's measured glucose level. **(a)** $P(X > 140) = P(Z > 1.5) = 0.0668$. **(b)** If $\bar{x}$ is the mean of three measurements (assumed to be independent), then $\bar{x}$ has a $N(125, 10/\sqrt{3})$ or $N(125$ mg/dl, 5.7735 mg/dl) distribution, and $P(\bar{x} > 140) = P(Z > 2.60) = 0.0047$.

5.27 The mean of three measurements has a $N(125$ mg/dl, 5.7735 mg/dl) distribution, and $P(Z > 1.645) = 0.05$ if Z is $N(0, 1)$, so $L = 125 + 1.645(5.7735) = 134.5$ mg/dl.

5.29 If W is total weight, and $\bar{x} = W/25$, then: $P(W > 5200) = P(\bar{x} > 208) = P\left(Z > \dfrac{208 - 190}{35 / \sqrt{25}} \right) = P(Z > 2.57) = 0.0051$.

5.31 (a) The mean of the difference (blue minus brown) will be $6.3 - 5.8 = 0.5$. The standard deviation will be $\sigma = \sqrt{0.283^2 + 0.174^2} = 0.332$. Because both sample means will be approximately Normal, the difference in the sample means will be approximately Normal. **(b)** We want $P(\bar{x}_{Brown} - \bar{x}_{Blue}) < 0$. This is $P\left(Z < \dfrac{0 - 0.5}{0.332} \right) = P(Z < -1.51)$. Table A gives us probability 0.0655. Software gives a probability of 0.0661.

5.33 (a) Assuming both samples are "large," $\overline{y}$ has a $N(\mu_Y, \sigma_Y / \sqrt{m})$ distribution and $\overline{x}$ has a $N(\mu_X, \sigma_X / \sqrt{n})$ distribution. **(b)** $\overline{y} - \overline{x}$ has a Normal distribution with mean $\mu_Y - \mu_X$ and standard deviation $\sqrt{\sigma_{\overline{y}}^2 + \sigma_{\overline{x}}^2}$.

5.35 $n = 1965$. $X = 0.48(1965) = 943$. $\hat{p} = 0.48$.

5.37 (a) $n = 1500$. **(b)** Answers and reasons will vary. **(c)** If the choice is "Yes," $X = 1025$. **(d)** For "Yes," $\hat{p} = 1025 / 1500 = 0.683$.

5.39 $B(10, 0.5)$.

5.41 (a) For the $B(6, 0.4)$ distribution, $P(X = 0) = 0.0467$ and $P(X \geq 4) = 0.1792$. **(b)** For the $B(6, 0.6)$ distribution, $P(X = 6) = 0.0467$ and $P(X \leq 2) = 0.1792$. **(c)** The number of "failures" in the $B(6, 0.4)$ distribution has the $B(6, 0.6)$ distribution. With six trials, 0 successes is equivalent to 6 failures, and 4 or more successes is equivalent to 2 or fewer failures.

5.43 From Exercise 5.42, we have $\mu_{\hat{p}} = 0.5$ and $\sigma_{\hat{p}} = \sqrt{\dfrac{(0.5)(1 - 0.5)}{200}} = 0.0354$. **(a)**

$$P(0.4 < \hat{p} < 0.6) = P\left(\frac{0.4 - 0.5}{0.0354} < z < \frac{0.6 - 0.5}{0.0354}\right) = P(-2.82 < z < 2.82).$$ Using Table A, this becomes $0.9976 - 0.0024 = 0.9952$. Using software, we find $P(0.4 < \hat{p} < 0.6) = 0.9953$. **(b)** Similarly, $P(0.45 < \hat{p} < 0.55) = P(-1.41 < z < 1.41) = 0.8415$. Using software gives 0.8422.

5.45 (a) $P(X = 5) = \dfrac{e^{-4} 4^5}{5!} = 0.1563$. **(b)** $P(X \leq 5) = \sum_{x=0}^{5} \dfrac{e^{-4} 4^x}{x!} = 0.7851$.

5.47 (a) Separate flips are independent (coins have no "memory," so they do not get on a "streak" of heads). **(b)** The coin is fair. The probabilities are still $P(H) = P(T) = 0.5$. **(c)** The parameters for a binomial distribution are n and p. $\hat{p}$ is a sample statistic. **(d)** This is best modeled with a Poisson distribution.

5.49 (a) A $B(200, p)$ distribution seems reasonable for this setting (even though we do not know what p is). **(b)** This setting is not binomial; there is no fixed value of n. **(c)** A $B(500, 1/12)$ distribution seems appropriate for this setting. **(d)** This is not binomial because separate cards are not independent.

5.51 (a) The distribution of those who say they have stolen something is $B(10, 0.2)$. The distribution of those who do not say they have stolen something is $B(10, 0.8)$. **(b)** X is the number who say they have stolen something. $P(X \geq 4) = 1 - P(X \leq 3) = 0.1209$.

5.53 (a) $\mu = 10(0.2) = 2$ will say they have stolen. $\mu = 8 = 10(0.8)$ will say they have not stolen.
(b) $\sigma = \sqrt{10(0.2)(1-0.2)} = 1.265$. **(c)** If $p = 0.1$, $\sigma = \sqrt{10(0.1)(1-0.1)} = 0.949$. If $p = 0.01$,
$\sigma = \sqrt{10(0.01)(1-0.01)} = 0.315$. As p gets smaller, the standard deviation becomes smaller.

5.55 (a) $P(X \le 7) = 0.0172$ and $P(X \le 8) = 0.0566$, so 7 is the largest value of m. **(b)** $P(X \le 5) = 0.0338$ and $P(X \le 6) = 0.0950$, so 5 is the largest value of m. **(c)** The probability will decrease. When the sample size increases, the mean will increase as well.

5.57 The count of 5s among n random digits has a binomial distribution with $p = 0.1$. **(a)** P(at least one 5) $= 1 - P$(no 5s) $= 1 - (0.9)^6 = 0.4686$. (Or take 0.5314 from Table C and subtract from 1.) **(b)** $\mu = (40)(0.1) = 4$.

5.59 (a) $n = 4$, $p = 0.7$. **(b)** See table below. Use software or the binomial formula, which here would be

$$P(X = x) = \binom{4}{x}(0.7)^x(1-0.7)^{4-x}.$$

(c) $\mu = 4(0.7) = 2.8$.

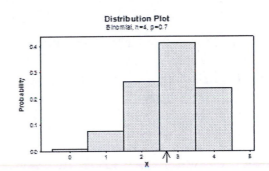

x	0	1	2	3	4
$P(x)$	0.0081	0.0756	0.2646	0.4116	0.2401

5.61 (a) Because $(0.7)(300) = 210$ and $(0.3)(300) = 90$, the approximate distribution is
$\hat{p} \sim N(0.7, \sqrt{\frac{(0.7)(0.3)}{300}} = 0.0265)$. $P(0.67 < \hat{p} < 0.73) = P\left(\frac{0.67 - 0.7}{0.0265} < z < \frac{0.73 - 0.7}{0.0265}\right) = P(-1.13 <$
$z < 1.13) = 0.8708 - 0.1292 = 0.7416$ (software gives 0.7424.) **(b)** If $p = 0.9$, the distribution of $\hat{p}$ is approximately $N(0.9, 0.0173)$ because $(300)(0.9) = 270$ and $(300)(0.1) = 30$.

$$P(0.87 < \hat{p} < 0.93) = P\left(\frac{0.87 - 0.9}{0.0173} < z < \frac{0.93 - 0.9}{0.0173}\right) = P(-1.73 < z < 1.73) = 0.9582 - 0.0418 =$$

0.9164 (software gives 0.9171.) **(c)** As p gets closer to 1, the probability of being within ± 0.03 of p increases because the standard deviation decreases.

5.63 (a) The mean is $\mu = p = 0.69$, and the standard deviation is $\sigma = \sqrt{p(1-p)/n} = 0.0008444$.
(b) $\mu \pm 2\sigma$ gives the range 68.83% to 69.17%. **(c)** This range is considerably narrower than the historical range. In fact, 67% and 70% correspond to $z = -23.7$ and $z = 11.8$, suggesting that the observed percents do not come from a $N(0.69, 0.0008444)$ distribution; that is, the population proportion has changed over time.

5.65 (a) $\hat{p} = 56/200 = 0.28$. **(b)** We need $P(\hat{p} \ge 0.28)$. $\hat{p}$ is approximately $N(0.24, 0.0302)$
because $(0.24)(200) = 48$ and $(0.76)(200) = 152$. $P(\hat{p} \ge 0.28) = P\left(z \ge \frac{0.28 - 0.24}{0.0302}\right) = P(Z \ge 1.32)$

$= 0.0934$. **(c)** Answers will vary. The key is that there is a decent probability (a bit less than 10%)

of obtaining a sample proportion 0.28 or larger if the proportion of binge drinkers at your campus is really 0.24.

5.67 (a) $p = 1/4 = 0.25$. **(b)** $P(X \geq 10) = 0.0139$. **(c)** $\mu = np = 5$ and $\sigma = \sqrt{np(1-p)} = \sqrt{3.75} = 1.9365$. **(d)** No. The trials would not be independent because the subject may alter his/her guessing strategy based on this information.

5.69 (a) X, the count of successes, has the $B(900, 1/5)$ distribution, with mean $\mu = np = (900)(1/5) = 180$ and $\sigma = \sqrt{(900)(0.2)(0.8)} = 12$ successes. **(b)** For $\hat{p}$, the mean is $\mu_{\hat{p}} = p = 0.2$ and $\sigma_{\hat{p}} = \sqrt{(0.2)(0.8)/900} = 0.01333$. **(c)** $P(\hat{p} > 0.24) = P(Z > 3) = 0.0013$. **(d)** From a standard Normal distribution, $P(Z > 2.326) = 0.01$, so the subject must score 2.326 standard deviations above the mean: $\mu_{\hat{p}} + 2.326\sigma_{\hat{p}} = 0.20 + 2.326(0.01333) = 0.2310$. This corresponds to 208 or more successes (correct guesses) in 900 attempts.

5.71 Jodi's number of correct answers will have the $B(n, 0.88)$ distribution. **(a)** $P(\hat{p} \leq 0.85) = P(X \leq 85)$ is on line 1. **(b)** $P(\hat{p} \leq 0.85) = P(X \leq 212)$ is on line 2. **(c)** For a test with 400 questions, the standard deviation of $\hat{p}$ would be half as big as the standard deviation of $\hat{p}$ for a test with 100 questions:

			With continuity correction	
Exact Prob.	Table Normal	Software Normal	Table Normal	Software Normal
0.2160	0.1788	0.1780	0.2206	0.2209
0.0755	0.0594	0.0597	0.0721	0.0722

With $n = 100$, $\sigma = \sqrt{(0.88)(0.12)/100} = 0.03250$; and with $n = 400$, $\sigma = \sqrt{(0.88)(0.12)/400} = 0.01625$. **(d)** Yes. Regardless of p, n must be quadrupled to cut the standard deviation in half.

5.73 Y has possible values 1, 2, 3, ... $P(\text{first one appears on toss } k) = (5/6)^{k-1}(1/6)$ because tossing a one must be preceded by $k - 1$ "not ones."

5.75 (a) $\mu = 50$. **(b)** The standard deviation is $\sigma = \sqrt{50} = 7.071$. $P(X > 60) = P(z > 1.41) = 0.0793$. Software gives 0.0786.

5.77 (a) With $\sigma_{\bar{x}} = 0.08/\sqrt{3} = 0.04619$, $\bar{x}$ has (approximately) a $N(123 \text{ mg}, 0.04619 \text{ mg})$ distribution. **(b)** $P(\bar{x} \geq 124) = P(Z \geq 21.65)$, which is essentially 0.

5.79 (a) The distribution will be approximately Normal with mean $\mu_{\bar{x}} = 2.13$ and standard deviation $\sigma_{\bar{x}} = 1.88/\sqrt{140} = 0.159$. **(b)** $P(\bar{x} < 2) = P(Z < -0.82) = 0.2061$. Software gives 0.2068. **(c)** Yes, because $n = 140$ is large.

5.81 The probability that the first digit is 1, 2, or 3 is $0.301 + 0.176 + 0.125 = 0.602$, so the number of invoices for amounts beginning with these digits should have a binomial distribution with $n = 1000$ and $p = 0.602$. More usefully, the proportion $\hat{p}$ of such invoices should have approximately a Normal distribution with mean $p = 0.602$ and standard deviation

$\sqrt{p(1-p)/1000} = 0.01548$, so $P(\hat{p} \le 560/1000 = 0.560) = P(Z \le -2.71) = 0.0034$.

5.83 If the carton weighs between 755 g and 830 g, then the average weight of the 12 eggs must be between $755/12 = 62.92$ g and $830/12 = 69.17$ g. The distribution of the mean weight is $N(66, 6/\sqrt{12} = 1.732)$. $P(62.92 < \bar{x} < 69.17) = P(-1.78 < Z < 1.83) = 0.9288$.

5.85 (a) He needs $\$520/\$35 = 14.857$ (really 15) wins. **(b)** $\mu = 520(0.026) = 13.52$ and $\sigma = \sqrt{520(0.026)(1-0.026)} = 3.629$. **(c)** Without the continuity correction, $P(X \ge 15) = P(Z \ge 0.41) = 0.3409$. With the continuity correction, we have $P(X \ge 14.5) = P(Z \ge 0.27) = 0.3936$. The continuity correction is much closer.

5.87 (a) $\hat{p}_F$ is approximately $N(0.82, 0.01921)$ and $\hat{p}_M$ is approximately $N(0.88, 0.01625)$. **(b)** When we subtract two independent Normal random variables, the difference is Normal. The new mean is the difference of the two means ($0.88 - 0.82 = 0.06$), and the new variance is the sum of the variances ($0.000369 + 0.000264 = 0.000633$), so $\hat{p}_M - \hat{p}_F$ is approximately $N(0.06, 0.02516)$. **(c)** $P(\hat{p}_F > \hat{p}_M) = P(\hat{p}_F - \hat{p}_M < 0) = P(Z < -2.38) = 0.0087$ (software gives 0.0085).

5.89 For each step of the random walk, the mean is $\mu = (1)(0.6) + (-1)(0.4) = 0.2$, the variance is $\sigma^2 = (1-0.2)^2(0.6) + (-1-0.2)^2(0.4) = 0.96$, and the standard deviation is $\sigma = \sqrt{0.96} = 0.9798$. Therefore, $Y/500$ has approximately a $N(0.2, 0.04382)$ distribution, and $P(Y \ge 200) = P(Y/500 \ge 0.4) \doteq P(Z \ge 4.56) = 0$.

    Note: *The number R of right-steps has a binomial distribution with n = 500 and p = 0.6. $Y \ge 200$ is equivalent to taking at least 350 right-steps, so we can also compute this probability as $P(R \ge 350)$, for which software gives the exact value 0.00000215. . . .*

Chapter 6 Solutions

6.1 $\sigma_{\bar{x}} = \sigma / \sqrt{36} = \$2.40/6 = \$0.40.$

6.3 Take two standard deviations: $0.80.

   Note: *This is the whole idea behind a confidence interval: Probability tells us that $\bar{x}$ is usually close to μ. That is equivalent to saying that μ is usually close to $\bar{x}$.*

6.5. (a) Results will vary. In one simulation, $23/25 = 92\%$ of the 80% confidence intervals contained μ. **(b)** Results will vary, but about 80% of the 80% confidence intervals should contain μ.

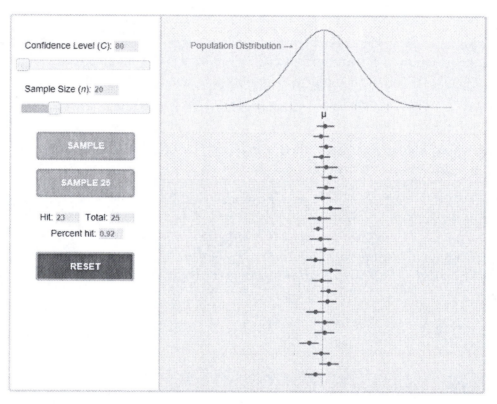

6.7 The margin of error would be halved because $n = 6400$ is roughly 4 times $n = 1601$. All else being equal, the width of the interval changes as $\sqrt{n}$. $\bar{x} \pm z^* \dfrac{\sigma}{\sqrt{n}} = 20,902 \pm 1.96 \dfrac{7500}{\sqrt{6400}} = 20,902 \pm 183.75$. The confidence interval would now be half as wide as in Exercise 6.6 because the new margin of error ($183.75) is half as large as the old margin of error ($367.39).

6.9 $n \geq \left(\dfrac{1.96 * 4300}{500} \right)^2 = 284.12$. Take $n = 285$. Note that you did not need the reported average salary; the margin of error for a confidence interval is based on the standard deviation, the confidence level, and the size of the sample.

6.11 The (useful) response rate is $532/1601 = 0.3323$, or about 33%. The reported margin of error is probably unreliable because we know nothing about the 67% of students that did *not* provide (useful) responses; they may be more (or less) likely to use credit cards. There are actually a whole host of possible reasons why a balance would not be given, so student responses will vary.

6.13. The margins of error are $(1.96)(28/\sqrt{n})$, which yields 17.355, 12.272, 8.677, and 6.136, respectively. (And, of course, all intervals are centered at 73.) Interval width decreases as sample size increases.

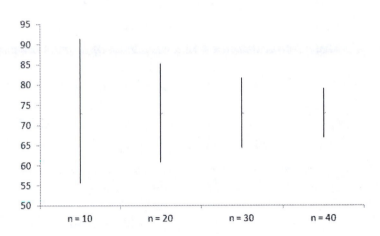

6.15 (a) She did not divide the standard deviation by $\sqrt{500} = 22.361$. **(b)** Confidence intervals concern the population mean, not the sample mean. (The value of the sample mean is known to be 8.6 and is the center of the interval; it is the population mean that we do not know.) **(c)** 95% is a confidence level, not a probability. Furthermore, it does not make sense to make probability statements about the population mean μ, once a sample has been taken and a confidence interval computed (at that point, both the interval and the unknown parameter are fixed and the interval either contains the parameter, or it does not). **(d)** The large sample size does not affect the distribution of individual alumni ratings (the population distribution). The use of a Normal distribution is justified because the distribution *of the sample mean* is approximately Normal when the sample is large, based on the central limit theorem.

    Note: *For part (c), a Bayesian statistician might view the population mean μ as a random quantity, but the viewpoint taken in this text is non-Bayesian.*

6.17 (a) The margin of error is $m = 1.96\dfrac{2.3}{\sqrt{340}} = 0.244$. The confidence interval is $\bar{x} \pm m$, or 5.156 to 5.644. **(b)** For 99% confidence, we have $m = 2.576\dfrac{2.3}{\sqrt{340}} = 0.321$. The 99% confidence interval becomes $5.4 \pm 0.321 = 5.079$ to 5.721. The 99% confidence margin of error is larger and the interval is wider.

6.19 For mean TRAP level, the margin of error is 2.29 U/l and the 95% confidence interval for μ is $13.2 \pm (1.96)(6.5/\sqrt{31}) = 13.2 \pm 2.29 = 10.91$ to 15.49 U/l.

6.21 Scenario A has a smaller margin of error. Both samples would have the same value of z^* (1.96), but the value of σ would likely be smaller for A because we might expect less variability in textbook cost for sophomore students than all students.

6.23 (a) The standard deviation of $\bar{x}$ is $\sigma_{\bar{x}} = 300/\sqrt{1000} = 9.49$. The probability is about 0.95 that $\bar{x}$ is within 18.98 kcal/day of . . . μ (because 18.98 is two standard deviations). **(b)** This is simply another way of understanding the statement from part (a): If $|\bar{x} - \mu|$ is less than 18.98 kcal/day 95% of the time, then "about 95% of all samples will capture the true mean . . . in the interval $\bar{x}$ plus or minus 18.98 kcal/day."

6.25 No. This is a range of values for the mean rent, not for individual rents.
     Note: *To find a range to include 95% of all rents, we should take $\mu \pm 2\sigma$ (or more precisely, $\mu \pm 1.96\sigma$), where μ is the (unknown) mean rent for all apartments, and σ is the standard deviation for all apartments (assumed to be \$220 in Exercise 6.24). If μ were equal to \$1050, for example, this range would be about \$618.80 to \$1481.20. However, because we do not actually know μ, we estimate it using $\bar{x}$, and to account for the variability in $\bar{x}$, we must widen the margin of error by a factor of $\sqrt{1 + 1/n}$. The formula $\bar{x} \pm 2\sigma\sqrt{1 + 1/10}$ is called a* prediction interval *for future observations. (Usually, such intervals are constructed with the* t *distribution, discussed in the Chapter 7, but the idea is the same.)*

6.27 (a) The 95% confidence interval for the mean number of hours spent listening to the radio in a week is $11.5 \pm (1.96)(8.3/\sqrt{1200}) = 11.5 \pm 0.470 = 11.03$ to 11.97 hours. **(b)** No. This is a range of values for the mean time spent, not for individual times. (See also the comment in the solution to Exercise 6.25.) **(c)** The large sample size ($n = 1200$ students surveyed) allows us to use the central limit theorem to say $\bar{x}$ has an approximately Normal distribution.

6.29 (a) We can be 95% confident, but not *certain*. (The true population percentage is either in the confidence interval, or it isn't.) **(b)** We obtained the interval 86.5% to 88.5% by a method that gives a correct result (that is, includes the true value of what we are trying to estimate) 95% of the time. **(c)** For 95% confidence, the margin of error is about two standard deviations (that is, $z^* = 1.96$), so $\sigma_{\text{estimate}} = 0.51\%$. **(d)** No, confidence intervals only account for random sampling error.

6.31 Multiply by $\left(\dfrac{1.609 \text{ km}}{1 \text{ mile}}\right)\left(\dfrac{1 \text{ gallon}}{3.785 \text{ liters}}\right) = 0.4251 \dfrac{\text{kpl}}{\text{mpg}}$. This gives $\bar{x}_{\text{kpl}} = 0.4251\bar{x}_{\text{mpg}} = 18.3515$ and margin of error $(1.96)(0.4251\sigma_{\text{mpg}}/\sqrt{20}) = 0.6521$ kpl, so the 95% confidence interval is 17.6994 to 19.0036 kpl.

6.33 $n \geq \left(\dfrac{1.96 * 6.5}{1.5}\right)^2 = 72.14$ —take $n = 73$.

6.35 No, this is not trustworthy. Because the numbers are based on voluntary response rather than an SRS, the confidence interval methods of this chapter cannot be used; the interval does not apply to the whole population.

6.37 The number of hits has a binomial distribution with parameters $n = 5$ and $p = 0.95$, so the number of misses is binomial with $n = 5$ and $p = 0.05$. We can therefore use Table C to answer these questions. **(a)** The probability that all cover their means is $0.95^5 = 0.7738$. (Or use Table C to find the probability of 0 misses.) **(b)** The probability that at least four intervals cover their means is $0.95^5 + 5(0.05)(0.95)^4 = 0.9774$. (Or use Table C to find the probability of 0 or 1 misses.)

6.39 If μ is the mean DXA reading for the phantom, we test H_0: $\mu = 1.4$ g/cm^2 versus H_a: $\mu \neq 1.4$ g/cm^2. We use the "not equal" alternate here because an "inaccurate" reading could deviate in either direction.

6.41 $P(Z < -1.81) = 0.03515$, so the two-sided P-value is $2(0.03515) = 0.0703$.

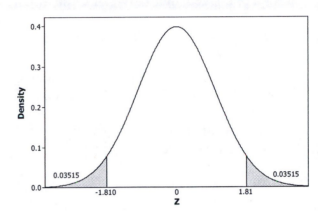

6.43 (a) For $P = 0.05$, the value of z is 1.645. **(b)** For a one-sided alternative (on the positive side), z is statistically significant at $\alpha = 0.05$ if $z > 1.645$.

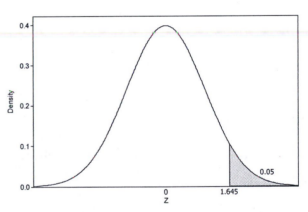

6.45 (a) $z = \dfrac{27.5 - 25}{8 / \sqrt{36}} = 1.875$. **(b)** For a one-sided alternative, $P = P(Z > 1.875) = 0.0304$. **(c)** For a two-sided alternative, double the one-sided P-value: $P = 0.0608$.

6.47 Recall the statement from the text: "A level α two-sided significance test rejects . . . H_0: $\mu = \mu_0$ exactly when the value μ_0 falls outside a level $1 - \alpha$ confidence interval for μ." **(a)** No. 30 is not in the 95% confidence interval because $P = 0.041$ means that we would reject H_0 at $\alpha = 0.05$. **(b)** Yes; 30 is in the 99% confidence interval because we would *not* reject H_0 at $\alpha = 0.01$.

6.49 (a) Yes, we reject H_0 at $\alpha = 0.05$ because $0.021 < 0.05$. **(b)** No, we do not reject H_0 at $\alpha = 0.01$ because $0.021 > 0.01$. **(c)** We have $P = 0.021$; we reject H_0 at significance level α if $P < \alpha$.

6.51 (a) One of the one-sided P-values is half as big as the two-sided P-value (0.031); the other is $1 - 0.031 = 0.969$. **(b)** Suppose the null hypothesis is H_0: $\mu = \mu_0$. The smaller P-value (0.031) goes with the one-sided alternative that is consistent with the observed data; for example, if $\bar{x} >$ μ_0, then $P = 0.031$ for the alternative $\mu > \mu_0$.

6.53 (a) Hypotheses should be stated in terms of the population mean, not the sample mean. **(b)** The null hypothesis H_0 should be that there is no change ($\mu = 21.2$). **(c)** A small P-value is needed for significance; $P = 0.98$ gives no reason to reject H_0. **(d)** We compare the P-value, not the z-statistic, to α. (In this case, such a small value of z would have a very large P-value—close to 0.5 for a one-sided alternative, or close to 1 for a two-sided alternative.)

6.55 (a) If μ is the mean score for the population of placement-test students, then we test H_0: $\mu = 77$ versus H_a: $\mu \neq 77$ because we have no prior belief about whether placement-test students will do better or worse (we just wanted to know if they differ). **(b)** If μ is the mean time to complete the maze with rap music playing, then we test H_0: $\mu = 20$ seconds versus H_a: $\mu > 20$ seconds because we believe rap music will make the mice finish more slowly. **(c)** If μ is the mean area of the apartments, we test H_0: $\mu = 880$ ft^2 versus H_a: $\mu < 880$ ft^2, because we suspect the apartments are smaller than advertised.

6.57 (a) H_0: $\mu = \$42,800$ versus H_a: $\mu > \$42,800$, where μ is the mean household income of mall shoppers. **(b)** H_0: $\mu = 0.4$ hr versus H_a: $\mu \neq 0.4$ hr, where μ is this year's mean response time.

6.59 (a) For H_a: $\mu > \mu_0$, the P-value is $P(Z > -1.69) = 0.9545$.
(b) For H_a: $\mu < \mu_0$, the P-value is $P(Z < -1.69) = 0.0455$.
(c) For H_a: $\mu \neq \mu_0$, the P-value is $2P(Z < -1.69) = 2(0.0455) = 0.0910$.

6.61 $P = 0.09$ means there is some evidence for the wage decrease, but it is not significant at the $\alpha = 0.05$ level. Specifically, the researchers observed that average wages for peer-driven students were 13% lower than average wages for ability-driven students, but (when considering overall variation in wages) such a difference might arise by chance 9% of the time, even if student motivation had no effect on wages.

6.63. Even if the two groups (the health and safety class and the statistics class) had the same level of alcohol awareness, there might be some difference in our sample due to chance. The difference observed was large enough that it would rarely arise by chance. The reason for this difference might be that health issues related to alcohol use are probably discussed in the health and safety class.

6.65 Because the difference for public school students was statistically significant, we can say the mean score for them increased. The difference for private school students was not significant; that does not mean that it didn't increase, but rather that it didn't increase *enough* to be called significant.

6.67 If μ is the mean difference between the two groups of children, we test H_0: $\mu = 0$ versus H_a: $\mu \neq 0$ (we only want to know if the difference is significant). The test statistic is
$z = \dfrac{5.8 - 0}{1.4} = 4.14$, for which software reports $P = 0.00003$—very strong evidence against the null hypothesis.

Note: *The exercise reports the standard deviation of the mean difference, rather than the sample standard deviation; that is, the reported value has already been divided by* $\sqrt{238}$.

6.69 If μ is the mean east-west location, the hypotheses are H_0: $\mu = 100$ versus H_a: $\mu \neq 100$ (as in the previous exercise). For testing these hypotheses, we find $z = \dfrac{113.8 - 100}{58 / \sqrt{584}} = 5.75$. This is highly significant ($P < 0.0001$), so we conclude that the trees are not uniformly spread from east to west.

6.71 (a) $z = \dfrac{127.8 - 115}{30 / \sqrt{25}} = 2.13,$ so the P-value is $P = P(Z > 2.13) = 0.0166$. This is strong evidence that the older students have a higher SSHA mean. **(b)** The important assumption is that this is an SRS from the population of older students. We also assume a Normal distribution, but this is not crucial provided there are no outliers and little skewness.

6.73 (a) H_0: $\mu = 0$ mpg versus H_a: $\mu \neq 0$ mpg, where μ is the mean difference. **(b)** The mean of the 20 differences is $\bar{x} = 2.73$, so $z = \dfrac{2.73 - 0}{3 / \sqrt{20}} = 4.07,$ for which $P < 0.0001$. We conclude that $\mu \neq 0$ mpg; that is, we have strong evidence that the computer's reported fuel efficiency differs from the driver's computed values.

6.75 For (b) and (c), either compare with the critical values in Table D or determine the P-value (0.0336). **(a)** H_0: $\mu = 0.61$ mg versus H_a: $\mu > 0.61$ mg. **(b)** Yes, because $z > 1.645$ (or because $P < 0.05$). **(c)** No, because $z < 2.326$ (or because $P > 0.01$).

6.77. See the sample screen (for $\bar{x} = 1$) at right. As one can judge from the shading under the Normal curve, it would take a value of at least about 0.7 to be significant, which is higher than in the previous exercise. (In fact, the cutoff is about 0.7355, which is approximately $2.326/\sqrt{10}$.) Smaller α means that $\bar{x}$ must be farther away from μ_0 in order to reject H_0.

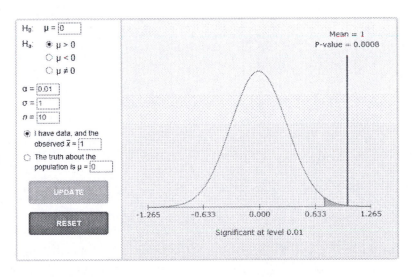

6.79 With sample size $n = 40$, sample means greater than 0.3 are statistically significant. See the result on the next page for $\bar{x} = 0.1$.

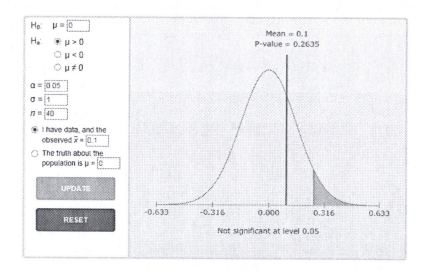

6.81 The *P*-values given by the applet are listed in the table; as $\overline{x}$ moves farther away from μ_0, *P* decreases. Note that these are (to within rounding) twice the *P*-values found in Exercise 6.80.

$\overline{x}$	P	$\overline{x}$	P
0.1	0.7518	0.6	0.0578
0.2	0.5271	0.7	0.0269
0.3	0.3428	0.8	0.0114
0.4	0.2059	0.9	0.0044
0.5	0.1139	1	0.0016

6.83 When a test is significant at the 1% level, it means that if the null hypothesis were true, outcomes similar to those seen are expected to occur fewer than 1 time in 100 repetitions of the experiment or sampling. "Significant at the 5% level" means we have observed something that occurs in fewer than 5 out of 100 repetitions (when H_0 is true). Something that occurs "fewer than 1 time in 100 repetitions" must also occur "fewer than 5 times in 100 repetitions," so significance at the 1% level guarantees significance at the 5% level (or any higher level). In mathematical terms, $P < 0.01$ means $P < 0.05$ as well.

6.85 Using Table D or software, we find that the 0.005 critical value is 2.576, and the 0.0025 critical value is 2.807. Therefore, if $2.576 < |z| < 2.807$—that is, either $2.576 < z < 2.807$ or $-2.807 < z < -2.576$—then z would be significant at the 1% level, but not at the 0.5% level with a two-sided test.

6.87 As $0.54 < 0.674$, the one-sided *P*-value is $P > 0.25$. (Software gives $P = 0.2946$.)

6.89 Because the alternative is two-sided, the answer for $z = -1.88$ is the same as for $z = 1.88$: $-1.645 > -1.88 > -1.960$, so Table D says that $0.05 < P < 0.10$, and Table A gives $P = 2(0.0301) = 0.0602$. Note there is no difference in the *P*-values for the two-sided alternative whether the sign on the test statistic is positive or negative.

6.91 In order to determine the effectiveness of alarm systems, we need to know the percent of all homes with alarm systems, and the percent of burglarized homes with alarm systems. For example, if only 10% of all homes have alarm systems, then we should compare the proportion of burglarized homes with alarm systems to 10%, not 50%. An alternate (but rather impractical) method would be to sample homes and classify them according to whether or not they had an alarm system, and also by whether or not they had experienced a break-in at some point in the recent past. This would likely require a very large sample in order to get a sufficiently large count of homes that had experienced break-ins.

6.93 The first test was barely significant at $\alpha = 0.05$, while the second was significant at any reasonable α.

6.95 A significance test answers only question (b). The P-value states how likely the observed effect (or a stronger one) is if H_0 is true, and chance alone accounts for deviations from what we expect. The observed effect may be significant (very unlikely to be due to chance) and yet not be of practical importance. And the calculation leading to significance *assumes* a properly designed study.

6.97 (a) If SES had no effect on LSAT results, there would still be some difference in scores due to chance variation. "Statistically insignificant" means that the observed difference was no more than we might expect from that chance variation. **(b)** If the results are based on a small sample, then even if the null hypothesis were not true, the test might not be sensitive enough to detect the effect. Knowing the effects were small tells us that the statistically insignificant test result did not occur merely because of a small sample size.

6.99 In each case, we find the test statistic z by dividing the observed difference $(2453.7 - 2403.7 = 50$ kcal/day) by $880/\sqrt{n}$. **(a)** For $n = 100$, $z = 0.57$, so $P = P(Z > 0.57) = 0.2843$. **(b)** For $n = 500$, $z = 1.27$, so $P = P(Z > 1.27) = 0.1020$. **(c)** For $n = 2500$, $z = 2.84$, so $P = P(Z > 2.84) = 0.0023$.

6.101 We expect more variation with small sample sizes than with large sample sizes, so even a large difference between $\bar{x}$ and μ_0 (or whatever measures are appropriate in our hypothesis test) might not turn out to be significant. If we were to repeat the test with a larger sample, the decrease in the standard error might give us a small enough P-value to reject H_0.

6.103 This exercise asks students to create their own examples. Answers will vary.

6.105 This exercise asks students to create their own examples. Answers will vary.

6.107 $P = 0.00001 = 1/100,000$, so we would need $n = 100,000$ tests in order to expect one P-value of this size (assuming that all null hypotheses are true). That is why we reject H_0 when we see P-values such as this: It indicates that our results would *very* rarely happen if H_0 were true.

6.109 Using $\alpha/12 = 0.05/12 = 0.004167$ as the cutoff, we reject the fifth ($P = 0.002$) and eleventh ($P < 0.002$) tests.

6.111 A larger sample gives more information and therefore gives a better chance of detecting a given alternative; that is, larger samples give more power.

6.113 The power for $\mu = 40$ will be higher than 0.6, because larger differences are easier to detect. The picture on the right shows one way to illustrate this (assuming Normal distributions): The solid curve (centered at 20) is the distribution under the null hypothesis, and the two dashed curves

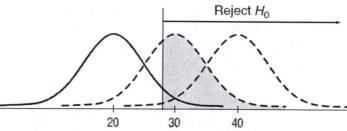

represent the alternatives $\mu = 30$ and $\mu = 40$. The shaded region under the middle curve is the power against $\mu = 30$; that is, that shaded region is 60% of the area under that curve. The power against $\mu = 40$ would be the corresponding area under the rightmost curve, which would clearly be greater than 0.6.

6.115 (a) Changing from the one-sided to the two-sided alternative decreases power (because more evidence is required to reject H_0; consider that the critical value for a one-sided H_a is 1.645, while it is 1.96 for a two-sided H_a). **(b)** Increasing σ decreases power (there is more variability in the population, so changes are harder to detect). **(c)** Power increases (larger sample sizes have narrower sampling distributions and hence more power to detect change).

6.117 The applet reports the power as 0.986.

6.119 We reject H_0 when $z > 2.326$, which is equivalent to $\bar{x} > 475 + 2.326 * 100 / \sqrt{500} = 485.4$, so the power against $\mu = 485$ is $P(\text{reject } H_0 \text{ when } \mu = 485) = P(\bar{x} > 485.4 \text{ when } \mu = 485) =$
$P\left(Z > \dfrac{485.4 - 485}{100 / \sqrt{500}}\right) = P(Z > 0.09) = 0.4641$. This is quite a bit less than the "80% power" standard.

6.121 (a) The hypotheses are "subject should go to college" and "subject should join work force." The two types of errors are recommending someone go to college when (s)he is better suited for the work force, and recommending the work force for someone who should go to college. **(b)** In significance testing, we typically wish to decrease the probability of wrongly rejecting H_0 (that is, we want α to be small); the answer to this question depends on which hypothesis is viewed as H_0.

   Note: *For part (a), there is no clear choice for which should be the null hypothesis. In the past, when fewer people went to college, one might have chosen "work force" as H_0—that is, one might have said, "we'll assume this student will join the work force unless we are convinced otherwise." Presently, roughly two-thirds of graduates attend college, which might suggest H_0 should be "college."*

6.123 From the description, we might surmise that we had two (or more) groups of students—say, an exercise group and a control (or no-exercise) group. **(a)** For example, if μ is the mean difference in scores between the two groups, we might test $H_0: \mu = 0$ versus $H_a: \mu \neq 0$. (Assuming we had no prior suspicion about the effect of exercise, the alternative should be two-sided.) **(b)** With $P = 0.27$, we would not reject H_0. In plain language: The results observed do not differ greatly from what we would expect if exercise had no effect on exam scores. **(c)** For example: Was this an experiment? What was the design? How big were the samples?

6.125 (a and b) The confidence intervals were computed as $\bar{x} \pm 1.96 s / \sqrt{n}$ and are shown in the table below. **(c)** Because the confidence interval for boys is entirely above the confidence interval for girls for each food intake, we could conclude that boys consume more of each, on average.

	Boys	Girls
Energy (kJ)	2399.9 to 2496.1	2130.7 to 2209.3
Protein (g)	24.00 to 25.00	21.66 to 22.54
Calcium (mg)	315.33 to 332.87	257.70 to 272.30

6.127 A sample screenshot and example plot are not shown but would be similar to those shown above for Exercise 6.126. Most students (99.4% of them) should find that their final proportion is between 0.84 and 0.96; 85% will have a proportion between 0.87 and 0.93.

6.129 (a) $\bar{x} = 5.3$ mg/dl, so $\bar{x} \pm 1.960 \sigma / \sqrt{6}$ is 4.61 to 6.05 mg/dl. **(b)** To test $H_0: \mu = 4.8$ mg/dl versus $H_a: \mu > 4.8$ mg/dl, we compute $z = \dfrac{\bar{x} - 4.8}{0.9 / \sqrt{6}} = 1.45$ and $P = 0.0735$. This is not strong enough to reject H_0.

    Note: *The confidence interval in (a) would allow us to say without further computation that, against a two-sided alternative, we would have P > 0.05. Because we have a one-sided alternative, we could conclude from the confidence interval that P > 0.025, but that is not enough information to draw a conclusion.*

6.131 (a) The stemplot is reasonably symmetric for such a small sample, but it could be viewed as skewed to the right. **(b)** $\bar{x} = 30.4$ μg/l; $30.4 \pm 1.96*7 / \sqrt{10}$ gives 26.06 to 34.74 μg/l. **(c)** We test $H_0: \mu = 25$ μg/l versus $H_a: \mu > 25$ μg/l. $z = \dfrac{30.4 - 25}{7 / \sqrt{10}} = 2.44$, so $P = 0.0073$. (We knew

```
2 034
2
3 01124
3 6
4 3
```

from (b) that it had to be smaller than 0.025 because 25 was not included in the 95% confidence interval). This is fairly strong evidence against H_0; the beginners' mean threshold is higher than 25 μg/l.

6.133 (a) Under H_0, $\bar{x}$ has a $N(0\%, 55\%/\sqrt{104}) = N(0\%, 5.3932\%)$ distribution.

(b) $z = \dfrac{6.9 - 0}{55 / \sqrt{104}} = 1.28$, so $P = P(Z > 1.28) = 0.1003$. **(c)** This is not significant at $\alpha = 0.05$. The study gives *some* evidence of increased compensation, but it is not very strong; similar results would happen about 10% of the time just by chance.

6.135 Yes. That's the heart of why we care about statistical significance. Significance tests allow us to discriminate between random differences ("chance variation") that might occur when the null hypothesis is true, and differences that are unlikely to occur when H_0 is true.

6.137 For each sample, find $\bar{x}$, then take $\bar{x} \pm (1.96)(4/\sqrt{12}) = \bar{x} \pm 2.2632$. We "expect" to see that 95 of the 100 intervals will include 25 (the true value of μ); binomial computations show that (about 99% of the time) 90 or more of the 100 intervals will include 20.

6.139 For each sample, find $\bar{x}$, then compute $z = \dfrac{\bar{x} - 23}{4/\sqrt{12}}$. Choose a significance level α and the appropriate cutoff point (z^*)—for example, with $\alpha = 0.10$, reject H_0 if $|z| > 1.645$; with $\alpha = 0.05$, reject H_0 if $|z| > 1.96$. Because the true mean is 25, $Z = \dfrac{\bar{x} - 25}{4/\sqrt{12}}$ has a $N(0, 1)$ distribution, so the probability that we will fail to reject H_0 is $P\left(-z^* < \dfrac{\bar{x} - 23}{4/\sqrt{12}} < z^*\right) = P\left(-z^* < Z + \dfrac{2}{4/\sqrt{12}} < z^*\right) = P(-1.7321 - z^* < Z < -1.7321 + z^*)$; here, we have adjusted for the difference between the true mean and the value of the mean in H_0. If $\alpha = 0.10$ $(z^* = 1.645)$, this probability is $P(-3.38 < Z < -0.09) = 0.4637$; if $\alpha = 0.05$ $(z^* = 1.96)$, this probability is $P(-3.69 < Z < 0.23) = 0.5909$. For smaller α, the probability will be larger. Thus, we "expect" to (wrongly) fail to reject H_0 about half the time (or more), and correctly reject H_0 about half the time or less. (The probability of rejecting H_0 is essentially the power of the test against the alternative $\mu = 25$.)

6.141 Answers will vary.

Chapter 7 Solutions

7.1. (a) The standard error of the mean is $s/\sqrt{n} = \$55/\sqrt{16} = \$13.75$. **(b)** The degrees of freedom are df $= n - 1 = 15$.

7.3 For the mean monthly rent, the 95% confidence interval for μ is $\$600 \pm (2.131)\$55/\sqrt{16} = \$600 \pm \$29.30 = \$570.70$ to $\$629.30$.

7.5. (a) Yes, $t = 2.22$ is significant when $n = 18$. This can be determined either by comparing to the df $= 17$ line in Table D (where we see that $t > 2.110$, the 2.5% critical value) or by computing the two-sided P-value (which is $P = 0.0403$). **(b)** No, $t = 2.22$ is not significant when $n = 9$, as can be seen by comparing to the df $= 8$ line in Table D (where we see that $t < 2.306$ the 2.5% critical value) or by computing the two-sided P-value (which is $P = 0.0572$). **(c)** Student sketches will likely be indistinguishable from

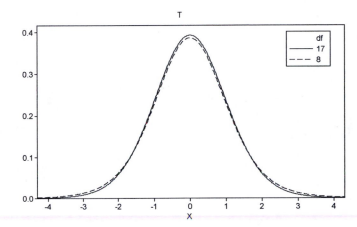

Normal distributions; careful students may try to show that the $t(8)$ distribution is shorter in the center and heavier to the left and right ("in the tails") than the $t(17)$ distribution (as is the case here), but in reality, the difference is nearly imperceptible.

7.7. Software will typically give a more accurate value for t^* than that given in Table D, and it will not round off intermediate values such as the standard deviation. Otherwise, the details of this computation are the same as what is shown in the textbook: df $= 7$, $t^* = 2.3646$, $7.1625 \pm t^*(3.5589/\sqrt{8}) = 7.1625 \pm 2.9753 = 4.1872$ to 10.1378, or about 4.19 to 10.14 hours per month. Output from Minitab follows.

One-Sample T: Watching Time

```
Variable        N  Mean  StDev  SE Mean     95% CI
Watching Time   8  7.16   3.56     1.26  (4.19, 10.14)
```

7.9. About 0.683 to 22.517: Using the mean and standard deviation from the previous exercise, the 95% confidence interval for μ is $11.6 \pm 2.7765(8.792/\sqrt{5}) = 11.6 \pm 10.9169 = 0.683$ to 22.517. (This is the interval produced by software; using the critical value $t^* = 2.776$ from Table D gives 0.685 to 22.515.)

7.11. The distribution is clearly non-Normal, but the sample size ($n = 63$) should be sufficient to overcome this, especially in the absence of strong skewness. One might question the independence of the observations; it seems likely that after 40 or so tickets had been posted for sale, that someone listing a ticket would look at those already posted for an idea of what price to charge. If we were to use *t* procedures, we would presumably take the viewpoint that these 63 observations come from a larger population of hypothetical tickets for this game, and we are trying to estimate the mean μ of that population. However, because (based on the histogram in Figure 1.31) the population distribution is likely bimodal, the mean μ might not be the most useful summary of this distribution.

7.13. As was found in Example 7.9, we reject H_0 if $t = \dfrac{\overline{x}}{1.5/\sqrt{20}} \geq 1.729$, which is equivalent to $\overline{x} \geq 0.580$. The power we seek is $P(\overline{x} \geq 0.580$ when $\mu = 1.05)$, which is

$$P\left(\frac{\overline{x} - 1.05}{1.5/\sqrt{20}} \geq \frac{0.580 - 1.05}{1.5/\sqrt{20}} \right) = P(Z \geq -1.40) = 0.9192.$$

7.15 With $n = 25$, we will reject H_0 if $t = \dfrac{\overline{x}}{1.5/\sqrt{25}} \geq 1.729$, which is equivalent to $\overline{x} \geq 0.5187$.

The power we seek is $P(\overline{x} \geq 0.5187$ when $\mu = 1)$, which is: $P\left(\dfrac{\overline{x} - 1}{1.5/\sqrt{25}} \geq \dfrac{0.5187 - 1}{1.5/\sqrt{25}} \right) =$

$P(Z \geq -1.60) = 0.9452$. The power found in Example 7.9 was 0.89.

7.17 (a) df = 11, $t^* = 2.201$. **(b)** df = 20, $t^* = 2.086$. **(c)** df = 20, $t^* = 1.725$. **(d)** For a given confidence level, t^* (and therefore the margin of error) decreases with increased sample size. For a given sample size, t^* increases with increased confidence.

7.19 The 5% critical value for a *t* distribution with df = 15 is 1.753. Only one of the one-sided options (reject H_0 when $t > 1.753$) is shown; the other is simply the mirror image of this sketch (shade the area to the left of -1.753, and reject when $t < -1.753$).

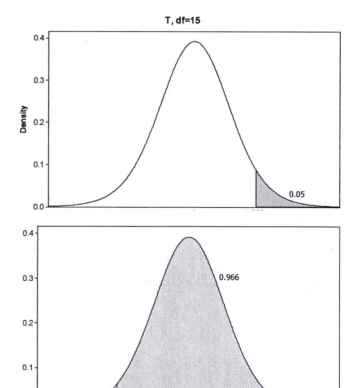

7.21 $\overline{x} = -11.2$ would support the alternative $\mu < 0$, and for that alternative, the *P*-value would still be 0.034. For the alternative $\mu > 0$ given in Exercise 7.20, the *P*-value is 0.966. Note that in the sketch shown, no scale has been given, because in the absence of a sample size, we do not know the degrees of freedom. Nevertheless, the *P*-value for the alternative $\mu > 0$ is the area above the computed value of the test statistic *t* (in the direction of the alternate hypothesis). As the area below t is 0.034, the area above this point must be 0.966.

7.23 (a) df = 26. **(b)** $1.706 < t < 2.056$. **(c)** Because the alternative is two-sided, we double the upper-tail probability to find the P-value: $0.05 < P < 0.10$. **(d)** $t = 2.01$ is not significant at either level (5% or 1%). **(e)** From software, $P = 0.0549$.

7.25 Let P be the given (two-sided) P-value, and suppose that the alternative is $\mu > \mu_0$. If $\bar{x}$ is greater than μ_0, this supports the alternative over H_0. However, if $\bar{x} < \mu_0$, we would not take this as evidence against H_0 because $\bar{x}$ is on the "wrong" side of μ_0. So, if the value of $\bar{x}$ is on the "correct" side of μ_0, the one-sided P-value is simply $P/2$. However, if the value of $\bar{x}$ is on the "wrong" side of μ_0, the one-sided P-value is $1 - P/2$ (which will always be at least 0.5, so it will never indicate significant evidence against H_0).

7.27 (a) If μ is the mean alcohol content by volume, we wish to test H_0: $\mu = 4.7\%$ versus H_a: $\mu \neq 4.7\%$. From these data, we have $\bar{x} = 4.9767$ and $s = 0.03215$. The test statistic is

$$t = \frac{4.9767 - 4.7}{0.03215 / \sqrt{3}} = 14.907.$$ With df = 2, we have $0.002 < P < 0.005$ (software gives $P = 0.0045$).

These data indicate the mean alcohol content of Budweiser beer is not 4.7%. **(b)** The 95% confidence interval is $4.9767 \pm 4.303(0.03215 / \sqrt{3}) = 4.9767 \pm 0.0799$, or 4.8968% to 5.0566%. Note that this interval is entirely above 4.7%. **(c)** For the cans and bottles to be within 0.3% of the advertised level, they need to be between 4.7% and 5.3%. Because our confidence interval is entirely contained in this range, it appears that Budweiser is within the standards.

7.29 (a) If μ is the mean number of uses a person can produce in 5 minutes after witnessing rudeness, we wish to test H_0: $\mu = 10$ versus H_a: $\mu < 10$. **(b)** $t = \dfrac{7.88 - 10}{2.35 / \sqrt{34}} = -5.2603$, with df = 33, for which $P < 0.0001$. This is very strong evidence that witnessing rudeness decreases performance.

7.31 (a) A stemplot (right) reveals that the distribution has two peaks and a high value (not quite an outlier). Both the stemplot and quantile plot show that the distribution is not Normal. The five-number summary is 2.2, 10.95, 28.5, 41.9, 69.3 (all in cm); a boxplot is not shown, but the long "whisker" between Q_3 and the maximum is an indication of skewness. **(b)** Maybe: We have a large enough sample to overcome the non-Normal distribution, but we are sampling from a small population. **(c)** The mean is $\bar{x} = 27.29$ cm, $s = 17.7058$ cm, and the margin of error is $t^* \cdot (s / \sqrt{40})$:

	0	222244
	0	579
	1	0113
	1	678
	2	2
	2	6679
	3	112
	3	5789
	4	0033444
	4	7
	5	112
	5	
	6	
	6	9

	df	t^*	Interval
Table D	30	2.042	$27.29 \pm 5.7167 = 21.57$ to 33.01 cm
Software	39	2.0227	$27.29 \pm 5.6626 = 21.63$ to 32.95 cm

(d) One could argue for either answer. We chose a random sample from this tract, so the main question is, can we view trees in this tract as being representative of similar trees in the same area?

7.33 (a) Yes, the sample size is large (and means are not necessarily values of the variable). **(b)** We wish to test H_0: $\mu = 0$ versus H_a: $\mu \neq 0$. Note that even though we might believe the return trip is perceived as shorter, we were asked if there is a *difference*. $t = \dfrac{-0.55 - 0}{2.16 / \sqrt{69}} = -2.115$ with df $= 68$ (or 60 with Table D). Using Table D, we have $0.02 < P < 0.04$, while software gives $P = 0.0381$. At the 0.05 level, we have evidence that the mean rating is different from 0.

7.35 We wish to test H_0: $\mu = 45$ versus H_a: $\mu > 45$. $t = \dfrac{52.98 - 45}{10.34 / \sqrt{50}} = 5.457$. Using df $= 49$, $P \approx 0$; with df $= 40$, $P < 0.0005$. These data do indicate that parents of ADHD children are extremely stressed (on average).

7.37 (a) We wish to test H_0: $\mu = 0$ versus H_a: $\mu \neq 0$, where μ is the mean change in NEAT. $t = \dfrac{328 - 0}{256 / \sqrt{16}} = 5.125$ with df $= 15$, for which $P = 0.00012$. There is strong evidence of a change in NEAT. **(b)** With $t^* = 2.131$, the 95% confidence interval for the change in NEAT activity is 191.6 to 464.4 kcal/day. This tells us how much of the additional calories might have been burned by the increase in NEAT: It consumed 19% to 46% of the extra 1000 kcal/day.

7.39 (a) We wish to test H_0: $\mu_c = \mu_d$ versus H_a: $\mu_c \neq \mu_d$, where μ_c is the mean computer-calculated mpg and μ_d is the mean mpg computed by the driver. Equivalently, we can state the hypotheses in terms of μ, the mean difference between computer- and driver-calculated mpgs, testing H_0: $\mu = 0$ versus H_a: $\mu \neq 0$. **(b)** With mean difference $\bar{x} = 2.73$ and standard deviation $s = 2.8015$, the test statistic is $t = \dfrac{2.73 - 0}{2.8015 / \sqrt{20}} = 4.358$ with df $= 19$, for which $P = 0.0003$. We have strong evidence that the results of the two computations are different.

7.41 (a) To test H_0: $\mu = 925$ picks versus H_a: $\mu > 925$ picks, we have $t = \dfrac{938.2 - 925}{24.2971 / \sqrt{36}} = 3.27$ with df $= 35$, for which $P = 0.0012$. **(b)** For H_0: $\mu = 935$ picks versus H_a: $\mu > 935$ picks, we have $t = \dfrac{938.2 - 935}{24.2971 / \sqrt{36}} = 0.80$, again with df $= 35$, for which $P = 0.2146$. **(c)** The 90% confidence interval from the previous exercise was 931.3 to 945.0 picks, which includes 935, but not 925. For a test of H_0: $\mu = \mu_0$ versus H_a: $\mu \neq \mu_0$, we know that $P < 0.10$ for values of μ_0 outside the interval, and $P > 0.10$ if μ_0 is inside the interval. The one-sided P-value would be half of the two-sided P-value.

7.43 (a) The differences are spread from -0.018 to 0.020 g, with mean $\bar{x} = -0.0015$ and standard deviation $s = 0.0122$ g. A stemplot is shown on the right; the sample is too small to make definitive judgments about

```
-1   85
-0   65
 0   255
 1
 2   0
```

skewness or symmetry, but there may be an outlier. **(b)** For $H_0: \mu = 0$ versus $H_a: \mu \neq 0$, we find

$$t = \frac{-0.0015 - 0}{0.0122 / \sqrt{8}} = -0.347 \text{ with df} = 7, \text{ for which } P = 0.7388.$$ We cannot reject H_0 based on this sample; the two operators are not significantly different in their results. **(c)** The 95% confidence interval for μ is $-0.0015 \pm 2.365 * 0.0122 / \sqrt{8} = -0.0015 \pm 0.0102 = -0.0117$ to 0.0087 g. **(d)** The subjects from this sample may be representative of future subjects, but the test results and confidence interval are suspect because this is not a random sample.

7.45 (a) We test $H_0: \mu = 0$ versus $H_a: \mu > 0$, where μ is the mean change in score (that is, the mean improvement, which is expected to increase). **(b)** The distribution is left-skewed, with mean $\bar{x} = 2.5$ and $s = 2.8928$.

```
-0   6
-0
-0
-0
 0   00011
 0   222333333
 0
 0   66666
```

(c) $t = \dfrac{2.5 - 0}{2.8928 / \sqrt{20}} = 3.8649$, df $= 19$, and $P = 0.0005$; there is strong evidence of improvement in listening test scores. **(d)** With df $= 19$, we have t^* $= 2.093$, so the 95% confidence interval is $2.5 \pm 2.093 \dfrac{2.898}{\sqrt{20}} = 1.15$ to 3.85.

7.47 We test H_0: median $= 0$ versus H_a: median > 0—or equivalently, $H_0: p = 1/2$ versus $H_a: p > 1/2$, where p is the probability that Jocko's estimate is higher. One difference is 0; of the nine non-zero differences, seven are positive. The P-value is $P(X \geq 7) = 0.0898$ from a $B(9, 0.5)$ distribution; there is not quite enough evidence to conclude that Jocko's estimates are higher. In Exercise 7.38 we were able to reject H_0; here we cannot.

Note: *The failure to reject H_0 in this case is because with the sign test, we pay attention only to the sign of each difference, not the size. In particular, the negative differences are each given the same "weight" as each positive difference, in spite of the fact that the negative differences are only $-\$50$ and $-\$55$, while most of the positive differences are larger. See the "Caution" about the sign test in your text.*

Minitab output: Sign Test of Median = 0 versus median > 0

```
 N   Below   Equal   Above      P   Median
10     2       1        7   0.0898  137.5
```

7.49 We test H_0: median $= 0$ versus H_a: median $\neq 0$. There were three negative and five positive differences, so the P-value is $2P(X \geq 5)$ for a binomial distribution with parameters $n = 8$ and $p = 0.5$. From Table C or software (Minitab output below), we have $P = 0.7266$, which supports H_0. The t test P-value in Exercise 7.44 was 0.6410.

Minitab output: Sign Test of Median = 0 versus median ≠ 0

```
 N   Below   Equal   Above      P   Median
 8     3       0        5   0.7266   3.500
```

7.51 We test H_0: median $= 0$ versus H_a: median > 0, or $H_0: p = 1/2$ versus $H_a: p > 1/2$. Out of the 20 differences, 17 are positive (and none equal 0). The P-value is $P(X \geq 17)$ for a $B(20, 0.5)$ distribution. From Table C or software (Minitab output below), we have $P = 0.0013$, so we reject

H_0 and conclude that the computer does record a higher gas mileage than the driver. (Using a t test in Exercise 7.39, we found $P = 0.0013$, which led to the same conclusion.)

Minitab output: Sign Test of Median = 0 versus median > 0

```
    N   Below   Equal   Above       P   Median
   20       3       0      17  0.0013    3.000
```

7.53 The standard deviation for the given data was $s = 0.012224$. With $\alpha = 0.05$, $t = \dfrac{\bar{x}}{s/\sqrt{15}}$, and df $= 14$, we reject H_0 if $|t| \geq 2.145$, which means $|\bar{x}| \geq (2.145)(s/\sqrt{15})$, or $|\bar{x}| \geq 0.00677$. Assuming $\mu = 0.002$: $P(|\bar{x}| \geq 0.00677) = 1 - P(-0.00677 \leq \bar{x} \leq 0.00677) =$

$$1 - P\left(\frac{-0.00677 - 0.002}{s/\sqrt{15}} \leq \frac{\bar{x} - 0.002}{s/\sqrt{15}} \leq \frac{0.00677 - 0.002}{s/\sqrt{15}} \right) = 1 - P(-2.78 \leq Z \leq 1.51)$$

$= 1 - (0.9345 - 0.0027) = 0.07$. The power is about 7% against this alternative—not surprising, given the small sample size, and the fact that the difference (0.002) is small relative to the standard deviation.

    Note: *Power calculations are often done with software. This may give answers that differ slightly from those found by the method described in the text. Most software does these computations with a "noncentral t distribution" (used in the text for two-sample power problems) rather than a Normal distribution, resulting in more accurate answers. In most situations, the practical conclusions drawn from the power computations are the same regardless of the method used.*

7.55 Taking $s = 1.5$ as in Example 7.9, the power for the alternative $\mu = 0.75$ is:

$$P\left(\bar{x} \geq \frac{t^* s}{\sqrt{n}} \text{ when } \mu = 0.75 \right) = P\left(\frac{\bar{x} - 0.75}{s/\sqrt{n}} \geq \frac{t^* s/\sqrt{n} - 0.75}{s/\sqrt{n}} \right) = P\left(Z \geq t^* - 0.5/\sqrt{n} \right).$$

Using trial-and-error, we find that with $n = 26$, power $= 0.7999$, and with $n = 27$, power $= 0.8139$. Therefore, we need $n > 26$.

7.57. We find $SE_D = \sqrt{\dfrac{8^2}{10} + \dfrac{12^2}{10}} = 4.5607$. The options for the 95% confidence interval for $\mu_1 - \mu_2$ are shown on the right. The instructions

df	t^*	Conf. interval
15.7	2.1236	-19.6851 to -0.3149
9	2.262	-20.3163 to 0.3163

for this exercise say to use the second approximation (df $= 9$), in which case we do not reject H_0, because 0 falls in the 95% confidence interval. Using the first approximation (df $= 15.7$, typically given by software), the interval is narrower, and we would reject H_0 at $\alpha = 0.05$ against a two-sided alternative. (In fact, $t = -2.193$, for which $0.05 < P < 0.1$ [Table D, df $= 9$], or $P = 0.0438$ [software, df $= 15.7$].)

7.59 Because $2.145 < t < 2.264$ and the alternative is two-sided, Table D tells us that the *P*-value is $0.04 < P < 0.05$. (Software gives $P = 0.0468$.) That is sufficient to reject H_0 at $\alpha = 0.05$.

7.61 SPSS and SAS give both results (the SAS output refers to the unpooled result as the Satterthwaite method), while JMP and Excel show only the unpooled procedures. The pooled *t* statistic is 2.279, for which $P = 0.052$.

　　Note: *When the sample sizes are equal—as in this case—the pooled and unpooled t statistics are equal. (See Exercise 7.62.) Both Excel and JMP refer to the unpooled test with the slightly-misleading phrase "assuming unequal variances." The SAS output also implies that the variances are unequal for this method. In fact, unpooled procedures make* no *assumptions about the variances. Finally, note that both Excel and JMP can do pooled procedures as well as the unpooled procedures that are shown.*

7.63 (a) Hypotheses should involve μ_1 and μ_2 (population means) rather than $\overline{x}_1$ and $\overline{x}_2$ (sample means). **(b)** The samples are not independent; we would need to compare the 56 males to the 44 females. **(c)** We need *P* to be small (for example, less than 0.10) to reject H_0. A large *P*-value like this gives no reason to doubt H_0. **(d)** Assuming the researcher computed the *t* statistic using $\overline{x}_1 - \overline{x}_2$, a positive value of *t* does not support H_a. (The one-sided *P*-value would be 0.982, not 0.018.)

7.65 (a) We cannot reject $H_0 : \mu_1 = \mu_2$ in favor of the two-sided alternative at the 5% level because $0.05 < P < 0.10$ (Table D) or $P \doteq 0.0771$ (software). **(b)** We could reject H_0 in favor of $H_a : \mu_1 < \mu_2$. A negative *t* statistic means that $\overline{x}_1 < \overline{x}_2$, which supports the claim that $\mu_1 < \mu_2$, and the one-sided *P*-value would be half of its value from part (a): $0.025 < P < 0.05$ (Table D) or $P = 0.0386$ (software).

7.67 The research question translates into hypotheses $H_0 : \mu_{Brown} = \mu_{Blue}$ and $H_a : \mu_{Brown} > \mu_{Blue}$.

The test statistic is $t = \dfrac{0.55 - (-0.38)}{\sqrt{1.68^2 / 40 + 1.53^2 / 40}} = 2.59$. With the computed df $= 77.33$, we have $P = 0.0058$. Using the conservative 39 (or 30 from Table D) df, we have $0.005 < P < 0.01$. In all cases, we conclude that H_0 should be rejected; brown-eyed people seem more trustworthy according to this experiment.

7.69 The nonresponse rate was $(3866 - 1839)/3866 = 0.5243$, or about 52.4% in spite of the incentive of the gift card raffle and repeated email reminders. We don't know if these students were not Facebook users, did not want to disclose their time spent on Facebook, their GPAs or other academic information, or just ignored the survey. Because of this, the results should be viewed with caution.

7.71 (a) Stemplots (right) do not look particularly Normal, but they have no extreme outliers or skewness, so *t* procedures should be reasonably safe. **(b)** The table of summary statistics is below on the left. **(c)** We wish to test $H_0 : \mu_N = \mu_S$ versus $H_a : \mu_N < \mu_S$. **(d)** We

Neutral		Sad	
0	0000000	0	0
0	55	0	5
1	000	1	000
1		1	555
2	00	2	0
		2	55
		3	00
		3	55
		4	00

find $SE_D = 0.3593$ and $t = -4.303$, so $P = 0.0001$ (df $= 26.5$) or $P < 0.0005$ (df $= 13$). Either way, we reject H_0. It appears sad people will spend more, on average. **(e)** The 95% confidence interval for the difference is one of the two options in the table below on the right.

Group	n	$\bar{x}$	s	df	t^*	Confidence interval
Neutral	14	$0.5714	$0.7300	26.5	2.0538	-2.284 to -0.808
Sad	17	$2.1176	$1.2441	13	2.160	-2.322 to -0.770

7.73 (a) Although the individual values are all integers between 1 and 5, inclusive, t procedures should be safe because the sample sizes are large. **(b)** Taco Bell: $\bar{x} = 4.1987$, $s = 0.8761$, $n = 307$. McDonald's: $\bar{x} = 3.9365$, $s = 0.8768$, $n = 362$. **(c)** We find

df	t^*	Confidence interval
649.4	1.9636	0.129 to 0.396
306	1.9677	0.128 to 0.391
100	1.984	0.127 to 0.397

$SE_D = 0.0680$ and $t = 3.85$, for which $P = 0.0001$ (df $= 649.4$) or $P < 0.0005$ (df $= 100$). Either way, there is strong evidence of a difference in opinions about service. **(d)** The 95% confidence interval for the difference is one of the three options in the table on the right—roughly 0.13 to 0.40.

7.75 (a) Assuming we have SRSs from each population, this seems reasonable.
(b) $H_0 : \mu_{Early} = \mu_{Late}$ and $H_a : \mu_{Early} \neq \mu_{Late}$. **(c)** $SE_D = 1.0534$, $t = 1.614$, $P = 0.1075$ (df $= 347.4$) or $P = 0.1081$ (df $= 199$). In either case, we fail to detect a difference in mean fat consumption for the two groups. **(d)** The 95% confidence interval for the difference in mean fat consumption is one of the three options in the table on the right. We note that all the intervals contain 0; these support the results of the test.

df	t^*	Confidence interval
347.4	1.9636	-0.372 to 3.772
199	1.9720	-0.377 to 3.777
100	1.984	-0.390 to 3.790

7.77 (a) This may be near enough to an SRS, if this company's working conditions were similar to that of other workers. **(b)** $SE_D = 0.7626$; regardless of how we choose df, the interval rounds to 9.99 to 13.01 mg·y/m^3.
(c) A one-sided alternative would seem to be reasonable

df	t^*	Confidence interval
137.1	1.9774	9.992 to 13.008
114	1.9810	9.989 to 13.011
100	1.984	9.987 to 13.013

here; specifically, we would likely expect that the mean exposure for outdoor workers would be lower. For testing H_0, we find $t = 15.08$, for which $P < 0.0001$ with either df $= 137$ or 114 (and for either a one- or a two-sided alternative). We have strong evidence that outdoor concrete workers have lower dust exposure than the tunnel workers. **(d)** The sample sizes are large enough that skewness should not matter.

7.79 To find a confidence interval $(\bar{x}_1 - \bar{x}_2) \pm t^* SE_D$, we need one of the following:
- Sample sizes and standard deviations, in which case we could find the interval in the usual way
- t and df because $t = (\bar{x}_1 - \bar{x}_2) / SE_D$, so we could compute $SE_D = (\bar{x}_1 - \bar{x}_2) / t$ and use df to find t^*
- df and a more accurate P-value from which we could determine t, and then proceed as above. The confidence interval could give us useful information about the magnitude of the difference (although with such a small P-value, we do know that a 95% confidence interval would not include 0).

7.81 This is a matched pairs design; for example, Monday hits are (at least potentially) not independent of one another. The correct approach would be to use one-sample t methods on the seven differences (Monday hits for design 1 minus Monday hits for design 2, Tuesday/1 minus Tuesday/2, and so on).

7.83 The next 10 employees who need screens might not be an independent group—perhaps they all come from the same department, for example. Randomization reduces the chance that we end up with such unwanted groupings.

7.85 (a) Stemplots, boxplots, and five-number summaries (in cm) are shown on the right. The north distribution is right-skewed, while the south distribution is left-skewed.

	Min	Q_1	M	Q_3	Max
North	2.2	10.2	17.05	39.1	58.8
South	2.6	26.1	37.70	44.6	52.9

North		South
43322	0	2
65	0	57
443310	1	2
955	1	8
	2	13
8755	2	689
0	3	2
996	3	566789
43	4	003444
6	4	578
4	5	0112
85	5	

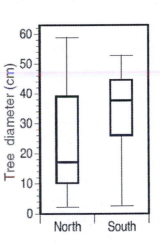

df	t^*	Confidence interval
55.7	2.004	−19.0902 to −2.5765
29	2.045	−19.2614 to −2.4053

(b) The methods of this section seem to be appropriate in spite of the skewness because the sample sizes are relatively large, and there are no outliers in either distribution. **(c)** We test $H_0 : \mu_N = \mu_S$ versus $H_a : \mu_N \neq \mu_S$; we should use a two-sided alternative because we have no reason (before looking at the data) to expect a difference in a particular direction. **(d)** The means and standard deviations are $\bar{x}_N = 23.7$, $s_N = 17.5001$, $\bar{x}_S = 34.53$, and $s_S = 14.2583$ cm. Then, $SE_D = 4.1213$, and $t = -2.629$ with df = 55.7 ($P = 0.011$) or df = 29 ($P = 0.014$). We conclude that the means are different (specifically, the south mean is greater than the north mean). **(e)** See the table for possible 95% confidence intervals. These intervals not only tell us that a difference exists, but that the northern trees are, on average, between about 2.5 and 19 cm smaller in dbh than the trees in the southern part of the tract.

7.87 (a) $SE_D = 1.9686$. Answers will vary with the df used (see the table), but the interval is roughly -1 to 7 units. **(b)** Because of random fluctuations between stores, we might (just by chance) have seen a rise in the average number of units sold even if actual mean sales

df	t^*	Confidence interval
122.5	1.9795	-0.90 to 6.90
54	2.0049	-0.95 to 6.95
50	2.009	-0.95 to 6.95

had remained unchanged. (Based on the confidence interval, mean sales might have even dropped slightly.)

7.89 (a) We test $H_0 : \mu_B = \mu_F$; $H_a : \mu_B > \mu_F$. $SE_D = 0.5442$ and $t = 1.654$, for which $P = 0.0532$ (df $= 37.6$) or 0.0577 (df $= 18$); there is not quite enough evidence to reject H_0 at $\alpha = 0.05$. **(b)** The confidence interval depends

df	t^*	Confidence interval
37.6	2.0251	-0.2021 to 2.0021
18	2.101	-0.2434 to 2.0434

on the degrees of freedom used; see the table. **(c)** We need two independent SRSs from Normal populations.

7.91 The pooled standard deviation is $s_p = 0.9347$, so the pooled standard error is

$s_p\sqrt{1/22 + 1/20} = 0.2888$. The test statistic is $t = 3.636$ with df $= 40$, for which $P = 0.0008$, and the 95% confidence interval (with $t^* = 2.021$) is 0.4663 to 1.6337. In Exercise 7.72, we reached the same conclusion on the significance test ($t = 3.632$ and $P = 0.0008$), and the confidence interval (using the more-accurate df $= 39.5$) was quite similar: 0.4655 to 1.6345.

7.93 See the solution to Exercise 7.85 for means and standard deviations. The pooled standard deviation is $s_p = 15.9617$, and the standard error is $SE_D = 4.1213$. The significance test has $t = -2.629$, df $= 58$, and $P = 0.0110$, so we have fairly strong evidence (though not quite significant at $\alpha = 0.01$) that the south mean is greater than the north mean. Possible answers for the confidence interval (with software, and with Table D) are given in the table. All results are similar to those found in Exercise 7.85.

df	t^*	Confidence interval
58	2.0017	-19.083 to -2.584
50	2.009	-19.113 to -2.554

Note: *If $n_1 = n_2$ (as in this case), the standard error and* t *statistic are the same for the usual and pooled procedures. The degrees of freedom will usually be different (specifically, df is larger for the pooled procedure, unless $s_1 = s_2$ and $n_1 = n_2$).*

7.95 With $s_N = 17.5001$, $s_S = 14.2583$, and $n_N = n_S = 30$, we have

$$df = \frac{(s_1^2 / n_1 + s_2^2 / n_2)^2}{(s_1^2 / n_1)^2 /(n_1 - 1) + (s_2^2 / n_2)^2 /(n_2 - 1)} = \frac{(17.5001^2 / 30 + 14.2583^2 / 30)^2}{(17.5001^2 / 30)^2 /(30 - 1) + (14.2582^2 / 30)^2 /(30 - 1)} = 55.725.$$

7.97 (a) With $s_I = 7.8$, $n_I = 115$, $s_O = 3.4$, and $n_O = 220$, we have

$$df = \frac{(7.8^2 / 115 + 3.4^2 / 220)^2}{(7.8^2 / 115)^2 /(115 - 1) + (3.4^2 / 220)^2 /(220 - 1)} = 137.066. \text{ This agrees within rounding.}$$

(b) $s_p = \sqrt{\dfrac{(n_I - 1)s_I^2 + (n_O - 1)s_O^2}{n_I + n_O - 2}} = 5.332$, which is slightly closer to s_O (the standard deviation from

the larger sample). **(c)** With no assumption of equality, $SE_D = \sqrt{s_I^2 / n_I + s_O^2 / n_O} = 0.7626$.

With the pooled method, $SE_D = s_p \sqrt{1/n_I + 1/n_O} = 0.6136$. **(d)** With the pooled standard

deviation, $t = 18.74$ and df $= 333$, for which $P < 0.0001$, and the 95% confidence interval is as shown in the table. With the smaller standard error, the t value is larger (it had been 15.08), and the confidence interval is narrower. The P-value is also smaller (although both are less than 0.0001). **(e)** With $s_I = 2.8$, $n_I = 115$, $s_O = 0.7$, and $n_O = 220$, we have

$$df = \frac{(2.8^2/115 + 0.7^2/220)^2}{(2.8^2/115)^2/(115 - 1) + (0.7^2/220)^2/(220 - 1)} = 121.503.$$ The pooled standard deviation is

$s_p = 1.734$; the standard errors are (with no assumptions) $SE = 0.2653$ and (assuming equal standard deviations) $SE = 0.1995$. The pooled t is 24.56 (df $= 333$, $P < 0.0001$), and the 95% confidence intervals are shown in the table. The pooled and usual t procedures compare similarly to the results for part (d): With the pooled procedure, t is larger, and the interval is narrower.

	df	t^*	Confidence interval
Part (d)	333	1.9671	10.2931 to 12.7069
	100	1.984	10.2827 to 12.7173
Part (e)	333	1.9671	4.5075 to 5.2925
	100	1.984	4.5042 to 5.2958

7.99 (a) From an $F(12, 21)$ distribution with $\alpha = 0.05$, $F^* = 2.25$. **(b)** Because $F = 2.45$ is greater than the 5% critical value, but less than the 2.5% critical value ($F^* = 2.64$), we know that P is between $2(0.025) = 0.05$ and $2(0.05) = 0.10$. (Software tells us that $P = 0.0697$.) $F = 2.45$ is significant at the 10% level but not at the 5% level.

7.101 The power would be higher. A smaller value of σ means that large differences between the sample means would not arise often by chance so that, if we observe such a difference, it gives more evidence of a difference in the population means.

    Note: *The table on the right shows the decrease in the power as σ increases.*

σ	Power
7.1	0.9105
7.2	0.9028
7.3	0.8950
7.4	0.7965
7.5	0.7844

7.103 $F = 6.1^2/5.8^2 = 1.106$. The degrees of freedom are 199 and 201. Using Table D (df $= 120$ and 200), $P > 0.200$. (Software gives $P = 0.4762$.)

7.105 The test statistic is $F = 7.8^2 / 3.7^2 = 5.2630$ with degrees of freedom 114 and 219. Table E tells us that $P < 0.002$, while software gives $P < 0.0001$; we have strong evidence that the standard deviations differ. The authors described the distributions as somewhat skewed, so the Normality assumption may be violated.

7.107 The test statistic is $F = 17.5001^2 / 14.2583^2 = 1.506$ with degrees of freedom 29 and 29. Table E tells us that $P > 0.2$, while software gives $P = 0.2757$; we cannot conclude that the standard deviations differ. The stemplots and boxplots of the north/south distributions in

Exercise 7.85 do not appear to be Normal (both distributions were skewed), so the results may not be reliable.

7.109 (a) To test $H_0 : \sigma_1 = \sigma_2$ versus $H_a : \sigma_1 \neq \sigma_2$, we find $F = 4.6224^2/4.3062^2 =$
1.152. We do not reject H_0. **(b)** With an $F(4, 4)$ distribution with a two-sided
alternative, we need the critical value for $p = 0.025$: $F^* = 9.60$. The table on the
right gives the critical values for other sample sizes. With such small samples, this
is a very low-power test; large differences between σ_1 and σ_2 would rarely be
detected.

n	F^*
5	9.60
4	15.44
3	39.00
2	647.79

7.111 The four standard deviations from Exercises 7.85 and 7.86 are $s_N = 17.5001$, $s_S = 14.2583$,
$s_E = 16.0743$, and $s_W = 15.3314$ cm. Using a larger σ for planning the study is advisable because
it provides a conservative (safe) estimate of the power. For example, if
we choose a sample size to provide 80% power and the true σ is smaller
than that used for planning, the actual power of the test is greater than the
desired 80%. Results of additional power computations depend on what
students consider to be "other reasonable values of σ." Shown in the
table are some possible answers using the Normal approximation.
(Powers computed using the noncentral t distribution are slightly
greater.)

	Power with $n =$	
σ	20	60
15	0.5334	0.9527
16	0.4809	0.9255
17	0.4348	0.8928
18	0.3945	0.8560

7.113 The mean is $\bar{x} = 139.5$, the standard deviation is $s = 15.0222$, and the standard error of the
mean is $s_{\bar{x}} = 7.511$. It would not be appropriate to construct a confidence interval because we
cannot consider these four scores to be an SRS.

7.115 The plot shows that t^* approaches 1.96 as df
increase.

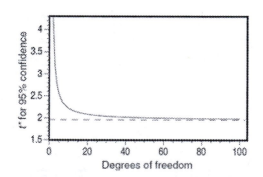

7.117 The margin of error is $t^*/\sqrt{n}$, using t^* for df $= n - 1$ and 95% confidence. For example,
when $n = 5$, the margin of error is 1.2417, and when $n = 10$, it is 0.7154, and for $n = 100$, it is
0.1984. As we see in the plot (below, left), as sample size increases, margin of error decreases
(toward 0, although it gets there very slowly).

7.119 (a) Use two independent samples (students who live in the dorms, and those who live
elsewhere). **(b)** Use a matched pairs design: Take a sample of college students, and have each
subject rate the appeal of each label design (make sure to randomize the order in which the
designs are presented). **(c)** Take a single sample of college students, and ask them to rate the
appeal of the product.

7.121 (a) To test $H_0 : \mu = 1.5$ versus $H_a : \mu < 1.5$, we have $t = \dfrac{0.83 - 1.5}{0.95 / \sqrt{200}} = -9.974$ with df = 199, for which $P \approx 0$. We can reject H_0 at the 5% significance level. **(b)** From Table D, use df = 100 and $t^* = 1.984$, so the 95% confidence interval for μ is $0.83 \pm 1.984(0.95/\sqrt{200}) = 0.83 \pm 0.133 = 0.697$ to 0.963 violations (with software, the interval is 0.698 to 0.962.) **(c)** While the significance test lets us conclude that there were less than 1.5 violations (on the average), the confidence interval gives us a range of reasonable values for the mean number of violations. **(d)** We have a large sample ($n = 200$), and the limited range means that there are no extreme outliers, so t procedures should be safe.

7.123 (a) The two-sample confidence interval (using the conservative df = 9) is $-0.853 \pm 2.262\sqrt{2.318^2 /10 + 1.924^2 /10} = -0.853 \pm 2.155 = -3.008$ to 1.302. (Software gives -2.859 to 1.153.) **(b)** The paired sample confidence interval is $-0.853 \pm 2.262(1.269 / \sqrt{10}) = -0.853 \pm 0.908 = -1.761$ to 0.055. **(c)** The centers of the intervals are the same, but the margin of error for the independent samples interval is much larger.

7.125 (a) The sample sizes are 41 and 197. We are looking at the average proportions across both samples. **(b)** Because we have no indication that a direction was considered prior to data being collected, the hypotheses are $H_0 : \mu_B = \mu_W$ and $H_a : \mu_B \neq \mu_W$, where μ_B is the average proportion of friends who are black for students with a black roommate and μ_W is the average proportion of friends who are black for students with a white roommate. **(c)** For the First Year: $t = 0.982$. With df = 52.3, $P = 0.3305$. With df = 40, $0.30 < P < 0.40$. For the Third Year: $t = 2.126$, df = 46.9, $P = 0.0388$. With df = 40, $0.02 < P < 0.04$. **(d)** There was not a significant difference in the average proportion of black friends at the middle of the first year; by the middle of the third year, the difference had become significant. Students with black roommates had a larger average proportion of black friends than those with white roommates.

7.127 (a) The mean difference in body weight change (with wine minus without wine) was $\bar{x}_1 = 0.4 - 1.1 = -0.7$ kg, with standard error $SE_1 = 8.6 / \sqrt{14} = 2.298$ kg. The mean difference in caloric intake was $\bar{x}_2 = 2589 - 2575 = 14$ cal, with $SE_2 = 210 / \sqrt{14} = 56.125$ cal. **(b)** The t statistics $t_i = \bar{x}_i / SE_i$, both with df = 13, are $t_1 = -0.305$ ($P_1 = 0.7655$) and $t_2 = 0.249$ ($P_2 = 0.8069$). **(c)** For df = 13, $t^* = 2.160$, so the 95% confidence intervals $\bar{x}_i \pm t^* SE_i$ are -5.66 to 4.26 kg (-5.67 to 4.27 with software) and -107.23 to 135.23 cal (-107.25 to 135.25 with software). **(d)** Students might note a number of factors in their discussions; for example, all subjects were males, weighing 68 to 91 kg (about 150 to 200 lb), which may limit how widely we can extend these conclusions.

7.129 (a) This is a matched pairs design because at each of the 24 nests, the same mockingbird responded on each day. **(b)** The variance of the difference is approximately $s_1^2 + s_4^2 - 2\rho s_1 s_4 = 48.684$, so the standard deviation is 6.9774 m. **(c)** To test $H_0 : \mu_1 = \mu_4$ versus $H_a : \mu_1 \neq \mu_4$, we have $t = \dfrac{15.1 - 6.1}{6.9774 / \sqrt{24}} = 6.319$ with df $= 23$, for which P is very small. **(d)** With correlation $\rho =$

0.3, the variance of the difference is approximately $s_1^2 + s_5^2 - 2\rho s_1 s_5 = 36.518$, so the standard

deviation is 6.0430 m. To test $H_0 : \mu_1 = \mu_5$ versus $H_a : \mu_1 \neq \mu_5$, we have

$t = \dfrac{4.9 - 6.1}{6.0430 / \sqrt{24}} = -0.973$ with df $= 23$, for which $P = 0.3407$. **(e)** The significant difference

between day 1 and day 4 suggests that the mockingbirds altered their behavior when approached by the same person for four consecutive days; seemingly, the birds perceived an escalating threat. When approached by a new person on day 5, the response was not significantly different from day 1; this suggests that the birds saw the new person as less threatening than a return visit from the first person.

7.131 How much a person eats or drinks may depend on how many people he or she is with. This means that the individual customers within each wine-label group probably cannot be considered to be independent of one another, which is a fundamental assumption of the t procedures.

7.133 No: What we have is nothing like an SRS of the population of school corporations; we have census data for your state.

7.135 The mean and standard deviation of the 25 numbers are $\bar{x} = 77.76\%$ and $s = 32.6768\%$, so the standard error is $SE_{\bar{x}} = 6.5354\%$. For df $= 24$, Table D gives $t^* = 2.064$, so the 95% confidence interval is $\bar{x} \pm 13.49\% = 64.27\%$ to 91.25%. This seems to support the retailer's claim: The original supplier's price was higher between 64% to 91% of the time.

7.137 Back-to-back stemplots follow. The distributions appear similar; the most striking difference is the relatively large number of boys with low GPAs. Testing the difference in GPAs ($H_0 : \mu_B = \mu_G$; $H_a : \mu_B < \mu_G$), we obtain $SE_D = 0.4582$ and $t = -0.91$, which is not significant, regardless of whether we use df $= 74.9$ ($P = 0.1839$) or 20 ($0.15 < P < 0.20$). The 95% confidence interval for the difference $\mu_B - \mu_G$ in GPAs is shown in the second table on the right. For the difference in IQs, we find $SE_D = 3.1138$. With the same hypotheses as before, we find $t = 1.64$—fairly strong evidence but not quite significant at the 5% level: $P = 0.0528$ (df $= 56.9$) or $0.05 < P < 0.10$ (df $= 30$). The 95% confidence interval for the difference $\mu_B - \mu_G$ in IQs is shown in the second table on the right.

		GPA		IQ	
	n	$\bar{x}$	s	$\bar{x}$	s
Boys	47	7.2816	2.3190	110.96	12.121
Girls	31	7.6966	1.7208	105.84	14.271

	df	t^*	Confidence interval
GPA	74.9	1.9922	-1.3277 to 0.4979
	30	2.042	-1.3505 to 0.5207
IQ	56.9	2.0025	-1.1167 to 11.3542
	30	2.042	-1.2397 to 11.4772

GPA:	Girls		Boys
		0	5
		1	7
		2	4
	4	3	689
	7	4	068
	952	5	0
	4200	6	019
	988855432	7	1124556666899
	998731	8	001112238
	95530	9	1113445567
	17	10	57

IQ:	Girls		Boys
	42	7	
		7	79
		8	
	96	8	
	31	9	03
	86	9	77
	433320	10	0234
	875	10	556667779
	44422211	11	00001123334
	98	11	556899
	0	12	03344
	8	12	67788
	20	13	
		13	6

7.139 It is reasonable to have a prior belief that people who evacuated their pets would score higher, so we test $H_0: \mu_1 = \mu_2$ versus $H_a: \mu_1 > \mu_2$. We find $SE_D = 0.4630$ and $t = 3.65$, which gives $P < 0.0005$ no matter how we choose degrees of freedom (115 or 237.0). As one might suspect, people who evacuated their pets have a higher mean score. One might also compute a 95% confidence interval for the difference; these are given in the table.

df	t^*	Confidence interval
237.0	1.9700	0.7779 to 2.6021
115	1.9808	0.7729 to 2.6071
100	1.984	0.7714 to 2.6086

7.141 The similarity of the sample standard deviations suggests that the population standard deviations are likely to be similar. The pooled standard deviation is $s_p = 436.368$ and $t = -0.3533$, so $P = 0.3621$ (df = 179), still not significant.

7.143 (a) We test $H_0: \mu_B = \mu_D$ versus $H_a: \mu_B < \mu_D$. Pooling might be appropriate for this problem, in which case $s_p = 6.5707$. Pooled or not, $SE_D = 1.9811$ and $t = 2.87$ with df = 42 (pooled), 39.3, or 21, so $P = 0.0032$, 0.0033, or 0.0046. We

	n	$\bar{x}$	s
Basal	22	41.0455	5.6356
DRTA	22	46.7273	7.3884
Strat	22	44.2727	5.7668

conclude that the mean score using DRTA is higher than the mean score with the Basal method. The difference in the average scores is 5.68; options for a 95% confidence interval for the difference $\mu_D - \mu_B$ are given in the table. **(b)** We test $H_0: \mu_B = \mu_S$ versus $H_a: \mu_B < \mu_S$. If we pool, $s_p = 5.7015$. Pooled or not, $SE_D = 1.7191$ and $t = 1.88$ with df = 42, 42.0, or 21, so $P = 0.0337$, 0.0337, or 0.0372. We conclude that the mean score using Strat is higher than the Basal mean score. The difference in the average scores is 3.23; options for a 95% confidence interval for the difference $\mu_S - \mu_B$ are given in the table.

df	t^*	Confidence interval for $\mu_D - \mu_B$
39.3	2.0223	1.6754 to 9.6882
21	2.0796	1.5618 to 9.8018
21	2.080	1.5610 to 9.8026
42	2.0181	1.6837 to 9.6799
40	2.021	1.6779 to 9.6857

df	t^*	Confidence interval for $\mu_S - \mu_B$
42.0	2.0181	−0.2420 to 6.6966
21	2.0796	−0.3477 to 6.8023
21	2.080	−0.3484 to 6.8030
42	2.0181	−0.2420 to 6.6965
40	2.021	−0.2470 to 6.7015

Chapter 8 Solutions

8.1 (a) $n = 5013$. **(b)** p is the (fixed, but unknown) population proportion of smartphone users who have purchased an item after using the phone to search for information. **(c)** $X = 2657$. X is the number of smartphone users from the sample who have purchased an item after using the phone to search for information. **(d)** $\hat{p} = 2657 / 5013 = 0.530$.

8.3 (a) $SE_{\hat{p}} = \sqrt{(0.530)(1-0.530)/5013} = 0.0070$. **(b)** The 95% confidence interval is $\hat{p} \pm 1.96 SE_{\hat{p}} = 0.530 \pm 1.96(0.0070) = 0.530 \pm 0.014$. **(c)** 51.6% to 54.4%.

8.5 For $z = 1.34$, the two-sided P-value is the area under a standard Normal curve above 1.34 and below -1.34.

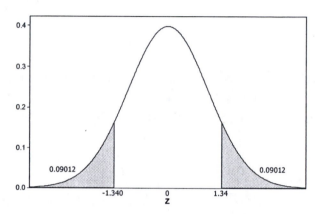

8.7 The sample proportion is $\hat{p} = 15/20 = 0.75$. To test $H_0: p = 0.5$ versus $H_a: p \neq 0.5$, the appropriate standard error is $\sigma_{\hat{p}} = \sqrt{p_0(1-p_0)/20} = 0.1118$, and the test statistic is $z = (0.75 - 0.5)/\sigma_{\hat{p}} = 0.250/0.1118 = 2.24$. The two-sided P-value is 0.0250 (Table A) or 0.0253 (software), so this result is significant at the 5% level.

8.9. (a) To test $H_0: p = 0.5$ versus $H_a: p \neq 0.5$ with $\hat{p} = 0.35$, the test statistic is
$$z = \frac{\hat{p} - p_0}{\sqrt{p_0(1-p_0)/n}} = -0.15/0.1118 = -1.34.$$ This is the opposite of the value of z given in Example 8.5, and the two-sided P-value is the same: 0.1802 (or 0.1797 with software). **(b)** The standard error for a confidence interval is $SE_{\hat{p}} = \sqrt{\hat{p}(1-\hat{p})/20} = 0.1067$, so the 95% confidence interval is $0.35 \pm 0.2090 = 0.1410$ to 0.5590. This is the complement of the interval shown in the Minitab output in Figure 8.3.

8.11. The plot is symmetric about 0.5, where it has its maximum. (The maximum margin of error always occurs at $\hat{p} = 0.5$, but the size of the maximum error depends on the sample size.)

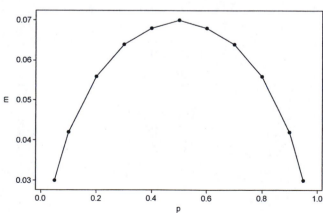

8.13 (a) $n = 200$, $X = 84$. **(b)** $\hat{p} = 84/200 = 0.42$. **(c)** $\hat{p} = 0.42$ (42%) is the estimate of p, the population proportion of students at your college who regularly eat breakfast.

8.15 (a) From Exercise 8.12, $\hat{p} = 0.461$, $SE_{\hat{p}} = \sqrt{0.461(1-0.461)/1003} = 0.0157$, $m = 1.96(0.0157) = 0.0308$. **(b)** There are certainly more than $20(1003) = 20,060$ cell phone owners; the number of "successes" was $462 > 10$ and the number of "failures" was $1003 - 462 = 541 > 10$. **(c)** $0.461 \pm 0.0308 = 0.4302$ to 0.4918. **(d)** We are 95% confident that between 43.0% and 49.2% of cell phone owners used their cell phone while in a store within the last 30 days to call a friend or family member for advice about a purchase they were considering.

8.17 (a) From Exercise 8.14, $\hat{p} = 180/230 = 0.7826$, $SE_{\hat{p}} = \sqrt{0.7826(1-0.7826)/230} = 0.0272$, $m = 1.96(0.0272) = 0.0533$. **(b)** The number of successes was 180, and the number of failures was $230 - 180 = 50$. Both counts are greater than 10. It is reasonable that more than $20(230) = 4600$ people are customers of the dealership, but this was not an SRS; they asked all customers in the two-week period. **(c)** The confidence interval is $0.7826 \pm 0.0533 = 0.7293$ to 0.8359. **(d)** Based on the sample, we estimate with 95% confidence that between 72.9% and 83.6% of the service department's customers would recommend the service to a friend; because this was not a SRS, we must use caution in relying on this interval.

8.19 Using the prior estimate $\hat{p} = 0.461$ as p^*, at 95% confidence,

$$n \geq \left(\frac{1.96}{0.04}\right)^2 (0.461)(1-0.461) = 596.6;$$ take at least $n = 597$ in the sample.

8.21 (a) The confidence level cannot exceed 100%. (In practical terms, the confidence level must be *less than* 100%.) **(b)** Margin of error only accounts for random sampling error. **(c)** P-values measure the strength of the evidence against H_0, not the probability of it being true.

8.23 The sample proportion is $\hat{p} = 3274/5000 = 0.6548$, the standard error is $SE_{\hat{p}} = 0.00672$, and the 95% confidence interval is $0.6548 \pm 0.0132 = 0.6416$ to 0.6680.

8.25 (a) $X = 0.89(1050) = 934.5$, which rounds to 935. We cannot have fractions of respondents. **(b)** $0.89 \pm 1.96\sqrt{0.89(1-0.89)/1050} = 0.8711$ to 0.9089. **(c)** 87.1% to 90.9%. **(d)** For example, parents might be conscious of violence because of recent events in the news.

8.27 (a) Values of $\hat{p}$ outside the interval

$$0.3 \pm 1.96\sqrt{\frac{0.3(1-0.3)}{50}} = 0.1730 \text{ to } 0.4270 \text{ will result}$$

in rejecting H_0. **(b)** For $n = 100$, values outside the interval 0.210 to 0.390 will result in rejecting H_0. **(c)** A sketch is included. What we see is the effect of the larger sample size; the distribution is narrowed, and so is the confidence interval.

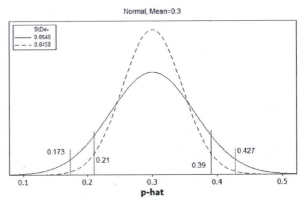

8.29 (a) About $(0.42)(159,949) = 67,179$ students plan to study abroad. **(b)** $SE_{\hat{p}} = 0.00123$, the margin of error is $(2.576)SE_{\hat{p}} = 0.00318$, and the 99% confidence interval is 0.4168 to 0.4232.

8.31 With $\hat{p} = 0.43$ and $n = 1430$, we have $SE_{\hat{p}} = 0.0131$, and the 95% confidence interval is $\hat{p} \pm 1.96\,SE_{\hat{p}} = 0.43 \pm 0.0257 = 0.4043$ to 0.4557.

8.33 (a) $SE_{\hat{p}} = \sqrt{(0.87)(0.13)/430,000} = 0.0005129$. For 99% confidence, the margin of error is $2.576\,SE_{\hat{p}} = 0.001321$. **(b)** One source of error is indicated by the wide variation in response rates: We cannot assume that the statements of respondents represent the opinions of nonrespondents. The effect of the participation fee is harder to predict, but one possible impact is on the types of institutions that participate in the survey: Even though the fee is scaled for institution size, larger institutions can more easily absorb it. These other sources of error are much more significant than sampling error, which is the only error accounted for in the margin of error from part (a).

8.35 (a) $\hat{p} = 390/1191 = 0.3275$. The standard error is $SE_{\hat{p}} = \sqrt{\hat{p}(1-\hat{p})/1191} = 0.01360$, so the margin of error for 95% confidence is $1.96\,SE_{\hat{p}} = 0.02665$, and the interval is 0.3008 to 0.3541. **(b)** Speakers and listeners probably perceive sermon length differently (just as, say, students and lecturers have different perceptions of the length of a class period).

8.37 (a) We test H_0: $p = 0.5$ versus H_a: $p > 0.5$, where p is the proportion who prefer fresh-brewed coffee. In the sample, 15 of 50 preferred instant, so $\hat{p} = 35/50 = 0.70$. The test statistic is

$$z = \frac{\hat{p} - p_0}{\sqrt{p_0(1-p_0)/n}} = \frac{0.7 - 0.5}{\sqrt{0.5*0.5/50}} = 2.83.$$

The *P*-value is 0.0023. **(b)** The graph is shown. **(c)** The test is significant at the 5% level (and the 1% level as well). We can conclude that people really do prefer fresh-brewed coffee over instant.

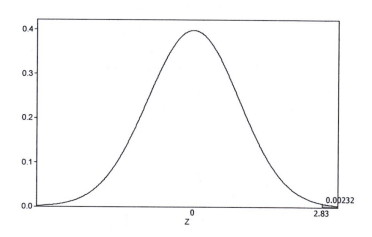

8.39. (a) For testing H_0: $p = 0.5$ versus H_a: $p \neq 0.5$, we have $\hat{p} = 5067/10,000 = 0.5067$

and $\sigma_{\hat{p}} = \sqrt{\dfrac{(0.5)(0.5)}{10,000}} = 0.005$, so

$z = \dfrac{0.0067}{0.005} = 1.34$, for which $P = 0.1802$.

This is not significant at $\alpha = 0.05$ (or even α

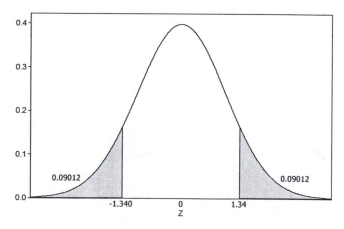

$= 0.10$). **(b)** $\text{SE}_{\hat{p}} = \sqrt{\dfrac{\hat{p}(1-\hat{p})}{10,000}} = 0.005$, so the 95% confidence interval is $0.5067 \pm (1.96)(0.005)$, or 0.4969 to 0.5165.

8.41. $n = \left(\dfrac{1.96}{0.01}\right)^2\left(\dfrac{1}{4}\right) = 9604$.

8.43. The required sample sizes are found by computing $\left(\dfrac{1.96}{0.08}\right)^2 p^*(1-p^*) = 600.25 p^*(1-p^*)$.

To be sure that we meet our target margin of error, we should take the largest sample indicated: $n = 151$ or larger. See the table and graph that follow.

p^*	n	Rounded up
0.1	54.0225	55
0.2	96.04	97
0.3	126.0525	127
0.4	144.06	145
0.5	150.0625	151
0.6	144.06	145
0.7	126.0525	127
0.8	96.04	97
0.9	54.0225	55

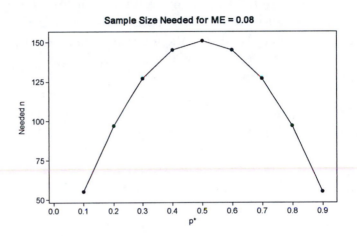

8.45. With $p_1 = 0.3$, $n_1 = 20$, $p_2 = 0.6$, and $n_2 = 30$, the mean and standard deviation of the sampling distribution of $D = \hat{p}_1 - \hat{p}_2$ are $\mu_D = p_1 - p_2 = -0.3$ and $\sigma_D = \sqrt{\dfrac{p_1(1-p_1)}{n_1} + \dfrac{p_2(1-p_2)}{n_2}} = 0.1360$.

8.47 (a) The means are $\mu_{\hat{p}_1} = p_1$ and $\mu_{\hat{p}_2} = p_2$. The standard deviations are $\sigma_{\hat{p}_1} = \sqrt{\dfrac{p_1(1-p_1)}{n_1}}$ and $\sigma_{\hat{p}_2} = \sqrt{\dfrac{p_2(1-p_2)}{n_2}}$. **(b)** $\mu_D = \mu_{\hat{p}_1} - \mu_{\hat{p}_2} = p_1 - p_2$. **(c)** $\sigma_D^2 = \sigma_{\hat{p}_1}^2 + \sigma_{\hat{p}_2}^2 = \dfrac{p_1(1-p_1)}{n_1} + \dfrac{p_2(1-p_2)}{n_2}$.

8.49 Let us call the proportions favoring Commercial B q_W and q_M. Our estimates of these proportions are the complements of those found in Exercise 8.48; for example, $\hat{q}_W = 56/100 = 0.56 = 1 - \hat{p}_W$. Consequently, the standard error of the difference $\hat{q}_W - \hat{q}_M$ is the same as that for $\hat{p}_M - \hat{p}_W$: $\text{SE}_D = \sqrt{\dfrac{\hat{q}_W(1-\hat{q}_W)}{100} + \dfrac{\hat{q}_M(1-\hat{q}_M)}{140}} = 0.06496$. The margin of error is therefore also the same, and the 95% confidence interval for $q_W - q_M$ is $(\hat{q}_w - \hat{q}_M) \pm (1.96)(0.06496) = -0.0030$ to 0.2516. Note that this is the same interval as that obtained in Exercise 8.48, but with the subtraction order switched.

   Note: *Here we followed the text's practice of subtracting the smaller proportion from the larger one.*

8.51. Because the sample proportions would tend to support the alternative hypothesis $(p_M > p_W)$, the P-value is half as large ($P = 0.0287$), which would be enough to reject H_0 at the 5% level.

8.53 (a) This was an experiment; the treatments were randomly assigned. Only 5 of 25 watched the second design for more than a minute; this does not fit the guidelines, as the count is less than 10. **(b)** It is reasonable to assume the sampled students were chosen randomly. No information was given about the size of the institution; are there more than $20(361) = 7220$ first-year students and more than $20(221) = 4420$ fourth-year students? There were more than 10 "yes" and "no" answers in each group.

8.55 (a) The number who watched (and did not watch) for more than one minute in each group was at least 5 (it was exactly 5 for those who watched the second version). **(b)** The number of "yes" and "no" answers in each group was more than 5.

8.57 (a) RR (watch more than one minute) $= (12/25)/(5/25) = 2.4$. **(b)** RR("yes" answer) $= 0.5294/0.2355 = 2.248$.

8.59 (a) Type of college is explanatory; response is whether physical education is required. **(b)** The populations are private and public colleges and universities. **(c)** $X_1 = 101$, $n_1 = 129$, $\hat{p}_1 = 101/129 = 0.7829$, $X_2 = 60$, $n_2 = 225$, $\hat{p}_2 = 0.2667$. **(d)**

$$(0.7829 - 0.2667) \pm 1.96\sqrt{\frac{0.7829(1-0.7829)}{129} + \frac{0.2667(1-0.2667)}{225}} = 0.5162 \pm 0.0917 = 0.4245 \text{ to}$$

0.6079. We note this interval does not contain 0; it appears that public institutions are more likely to require physical education. **(e)** We only want to compare, so the hypotheses are $H_0 : p_1 = p_2$ and $H_a : p_1 \neq p_2$. We have $\hat{p} = \dfrac{60+101}{225+129} = 0.4548$. The test statistic is

$$z = \frac{0.7829 - 0.2667}{\sqrt{0.4548(1-0.4548)\left(\dfrac{1}{225} + \dfrac{1}{129}\right)}} = 9.39, \; P \approx 0.$$ **(f)** There were 101 public institutions that

require physical education and 18 that do not. There were 60 private institutions that require physical education and 165 that do not. All these counts are greater than 5. We do not know if the samples were SRSs. **(g)** It appears that public institutions are much more likely to require physical education (by an estimated 42% to 61% at 95% confidence).

8.61 $SE_D = \sqrt{\dfrac{0.299(1-0.299)}{358} + \dfrac{0.208(1-0.208)}{851}} = 0.0279. \ (0.299-0.208) \pm 1.96(0.0279) = 0.0363$
to 0.1457. Based on these samples, we'd estimate at 95% confidence that among Canadian youth in 10th and 11th grades, between 3.6% and 14.6% more youths who stress about their health are exergamers than those who do not stress about their health.

8.63 (a) The filled-in table is on the right. The values of X_1 and X_2 are estimated as $(0.54)(1063)$ and $(0.89)(1064)$. **(b)** The estimated difference is $\hat{p}_2 - \hat{p}_1 = 0.35$.

Population	Population proportion	Sample size	Count of successes	Sample proportion
1 (adults)	p_1	1063	574	0.54
2 (teens)	p_2	1064	947	0.89

(c) Large-sample methods should be appropriate because we have large, independent samples from two populations, and we have at least 5 successes and failures in each sample. **(d)** With SE_D = 0.01805, the 95% confidence interval is $0.35 \pm 0.03537 = 0.3146$ to 0.3854. **(e)** The estimated difference is about 35%, and the interval is about 31.5% to 38.5%. **(f)** A possible concern is that adults were surveyed before Christmas, while teens were surveyed before and after Christmas. It might be that some of those teens may have received game consoles as gifts, but eventually grew tired of them.

8.65 (a) The filled-in table is on the right. The values of X_1 and X_2 are estimated as $(0.73)(1063)$ and $(0.76)(1064)$. **(b)** The estimated difference is $\hat{p}_2 - \hat{p}_1 = 0.03$.

Population	Population proportion	Sample size	Count of successes	Sample proportion
1 (adults)	p_1	1063	776	0.73
2 (teens)	p_2	1064	809	0.76

(c) Large-sample methods should be appropriate because we have large, independent samples from two populations, and we have at least 5 successes and failures in each sample. **(d)** With SE_D = 0.01889, the 95% confidence interval is $0.03 \pm 0.03702 = -0.0070$ to 0.0670. **(e)** The estimated difference is about 3%, and the interval is about -0.7% to 6.7%. **(f)** As in the solution to Exercise 8.63, a possible concern is that adults were surveyed before Christmas.

8.67 No; this procedure requires independent samples from different populations. We have one sample (of teens).

8.69 (a) H_0 should refer to p_1 and p_2 (population proportions) rather than $\hat{p}_1$ and $\hat{p}_2$ (sample proportions). **(b)** Knowing $\hat{p}_1 = \hat{p}_2$ does not tell us that the success counts are equal $(X_1 = X_2)$ *unless* the sample sizes are equal $(n_1 = n_2)$. **(c)** Confidence intervals only account for random sampling error.

8.71 (a) $\hat{p}_F = 48/60 = 0.8$, so $SE_{\hat{p}} = 0.05164$ for females. $\hat{p}_M = 52/132 = 0.3939$, so $SE_{\hat{p}} = 0.04253$ for males. **(b)** $SE_D = \sqrt{0.05164^2 + 0.04253^2} = 0.06690$, so the interval is $(\hat{p}_F - \hat{p}_M) \pm 1.645 SE_D$, or 0.2960 to 0.5161. There is (with high confidence) a considerably higher percent of juvenile references to females than to males. **(c)** $H_0 : p_F = p_M$ and

$H_a : p_F \neq p_M$. We have $\hat{p} = \dfrac{48 + 52}{60 + 132} = 0.5208$. The test statistic is

$$z = \frac{0.8 - 0.39}{\sqrt{0.5208(1 - 0.5208)\left(\dfrac{1}{60} + \dfrac{1}{132}\right)}} = 5.22, P \approx 0.$$ This test shows the difference in juvenile

references is statistically significant; we can observe that more female references are juvenile than are male references.

8.73 (a) $n = 2342$, $x = 1639$. **(b)** $\hat{p} = 1639 / 2342 = 0.6998$. SE $= \sqrt{0.6998(1 - 0.6998) / 2342} = 0.0095$. **(c)** $0.6998 \pm 1.96(0.0095) = 0.6812$ to 0.7184. We estimate at 95% confidence that between 68.1% and 71.8% of desktop and laptop computer owners get their news from their computer. **(d)** Yes. The counts of "successes" and "failures" are both larger than 15. We can assume Pew took a random sample that is approximately an SRS, and the population of computer owners is much larger than the sample size.

8.75 We have large samples from two independent populations (different age groups). $\hat{p}_1 = 861/1055 = 0.8161$, $\hat{p}_2 = 417/974 = 0.4281$. SE$_D = \sqrt{\dfrac{0.8161(0.1839)}{1055} + \dfrac{0.4281(0.5719)}{974}} = 0.0198$. The 95% confidence interval for the difference in the proportion of children in these age groups who get enough calcium in their diets is $(0.8161 - 0.4281) \pm 1.96(0.0198) = 0.3492$ to 0.4268. Because the interval is entirely above 0, we can conclude that children 5 to 10 years old are much more likely to get adequate calcium in their diets than children 11 to 13 years old.

8.77 Answers will vary. The confidence interval gives more information on the actual size of the difference.

8.79 (a) $0.67(1802) = 1207$. **(b)** $0.67 \pm 1.96(0.01108) = 0.6483$ to 0.6917. **(c)** About 64.8% to 69.2% of all Internet users use Facebook, at 95% confidence.

8.81 No. Many people use both; there was only one sample, not two independent samples.

8.83 (a) While there is only a 5% chance of any interval being wrong, we have six (roughly independent) chances to make that mistake. **(b)** For 99.2% confidence, use $z^* = 2.65$. (Using software, $z^* = 2.6521$, or 2.6383 using the exact value of 0.05/6). **(c)** The margin of error for each interval is $z^* \text{SE}_{\hat{p}}$, so each interval is about 1.35 times wider than in Exercise 8.82. (If intervals are rounded to three decimal places, as on the right, the results are the same regardless of the value of z^* used.)

Genre	Interval
Racing	0.705 to 0.775
Puzzle	0.684 to 0.756
Sports	0.643 to 0.717
Action	0.632 to 0.708
Adventure	0.622 to 0.698
Rhythm	0.571 to 0.649

8.85 The pooled estimate is $\hat{p} = 0.375$ (the average of $\hat{p}_1$ and $\hat{p}_2$, because the sample sizes were equal). Then, SE$_D = 0.01811$, so $z = (0.43 - 0.32)/\text{SE}_D = 6.08$, for which $P < 0.0001$.

8.87 With $\hat{p} = 1006/1530 = 0.6575$, we have $SE_{\hat{p}} = 0.01213$, so the 95% confidence interval is $\hat{p} \pm 1.96\ SE_{\hat{p}} = 0.6575 \pm 0.0238 = 0.6337$ to 0.6813.

8.89 We test $H_0 : p_F = p_M$ versus $H_a : p_F \neq p_M$. Finding the number of men and women who drink at least five servings per week of soft drinks, we have $X_M = (0.17)(1003) = 171$ and $X_F = (0.15)(1003) = 150$. $\hat{p} = 0.1600$, $SE_{D_p} = 0.0164$, $z = 1.28$, $P = 0.2009$. There is no significant difference between Australian men and women in terms of drinking five or more soft drinks per week.

8.91 This exercise asks students to find a Gallup poll; results will vary.

8.93 The proportions, z-values, and P-values are

Text	1	2	3	4	5	6	7	8	9	10
$\hat{p}$	.8718	.9000	.5372	.6738	.9348	.6875	.6429	.6471	.7097	.8759
z	4.64	6.69	0.82	5.31	5.90	5.20	3.02	2.10	6.60	9.05
P	≈ 0	≈ 0	0.4122	≈ 0	≈ 0	≈ 0	.0025	.0357	≈ 0	≈ 0

We reject $H_0: p = 0.5$ for all texts except Text 3 and (perhaps) Text 8. If we are using a "multiple comparisons" procedure such as Bonferroni (see Chapter 6), then we also might fail to reject H_0 for Text 7.

The last three texts do not seem to be any different from the first seven; the gender of the author does not seem to affect the proportion.

8.95 The difference becomes more significant (i.e., the P-value decreases) as the sample size increases. For small sample sizes, the difference between $\hat{p}_1 = 0.55$ and $\hat{p}_2 = 0.45$ is not significant, but with larger sample sizes, we expect that the sample proportions should be better estimates of their respective population proportions, so $\hat{p}_1 - \hat{p}_2 = 0.1$ suggests that $p_1 \neq p_2$.

n	z	P
60	1.10	0.2713
70	1.18	0.2380
80	1.26	0.2077
100	1.41	0.1585
400	2.83	0.0047
500	3.16	0.0016
1000	4.47	0.0000

8.97 (a) Using either trial and error, or the formula derived in part (b), we find that at least $n = 534$ is needed. **(b)** Generally, the margin of error is $m = z^* \sqrt{\dfrac{\hat{p}_1(1 - \hat{p}_1)}{n} + \dfrac{\hat{p}_2(1 - \hat{p}_2)}{n}}$; with $\hat{p}_1 = \hat{p}_2 = 0.5$, this is $m = z^* \sqrt{0.5/n}$. Solving for n, we find $n = (z^*/m)^2/2$.

8.99. (a) $p_0 = 143{,}611/181{,}535 = 0.7911$. **(b)** $\hat{p} = 339/870 = 0.3897$, $\sigma_{\hat{p}} = 0.0138$, and $z = (\hat{p} - p_0)/\sigma_{\hat{p}} = -29.1$, so $P = 0$ (regardless of whether H_a is $p < p_0$ or $p \neq p_0$). This is very strong evidence against H_0; we conclude that Mexican Americans are underrepresented on juries. **(c)** $\hat{p}_1 = 339/870 = 0.3897$, while $\hat{p}_2 = (143{,}611 - 339)/(181{,}535 - 870) = 0.7930$. Then, $\hat{p} = 0.7911$ (the value of p_0 from part (a)), $\mathrm{SE}_{D_p} = 0.01382$, and $z = -29.2$—and again, we have a tiny P-value and reject H_0.

8.101 In each case, the standard error is $\sqrt{\hat{p}(1-\hat{p})/1280}$. One observation is that, while many feel that loans are a burden and wish they had borrowed less, a majority are satisfied with the benefits they receive from their education.

	$\hat{p}$	$\mathrm{SE}_{\hat{p}}$	Interval
Burdened by debt	0.555	0.01389	0.5278 to 0.5822
Would borrow less	0.544	0.01392	0.5167 to 0.5713
More hardship	0.343	0.01327	0.3170 to 0.3690
Loans worth it	0.589	0.01375	0.5620 to 0.6160
Career opportunities	0.589	0.01375	0.5620 to 0.6160
Personal growth	0.715	0.01262	0.6903 to 0.7397

Chapter 9 Solutions

9.1 (a) Yes: 47/292 = 0.161, No: 245/292 = 0.839. **(b)** Yes: 21/233 = 0.090, No: 212/233 = 0.910. **(c)** See graph at right. **(d)** Females are somewhat more likely than males to have increased the time they spend on Facebook; however, the vast majority of both genders do not report increasing the time they spend on Facebook.

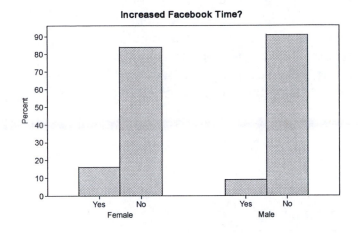

9.3 Answers will vary.

9.5 Among all three fruit consumption groups, vigorous exercise is most likely. Incidence of low exercise decreases with increasing fruit consumption.

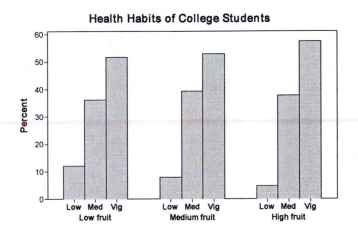

9.7 All expected counts are computed as $E_{ij} = \dfrac{(\text{row } i \text{ total})(\text{column } j \text{ total})}{\text{table total}}$. For example,

$$E_{low,low} = \frac{(569)(108)}{1184}.$$

	Physical Activity			
Fruit consumption	Low	Medium	Vigorous	Total
Low	51.9	212.9	304.2	569
Medium	29.3	120.1	171.6	321
High	26.8	110.0	157.2	294
Total	108	443	633	1184

9.9 (a) df = (5 − 1)(4 − 1) = 12. From Table F, 0.05 < P < 0.10. **(b)** df = 12, 0.05 < P < 0.10. **(c)** df = 1, 0.005 < P < 0.01. **(d)** df = 1, 0.20 < P < 0.25.

9.11 (a) The conditional distributions are given in the table. Divide the number of "yes" and "no" answers in each column by the column total. **(b)** The graph is shown at right. **(c)** Explanatory variable value 1 had proportionately fewer "yes" responses.

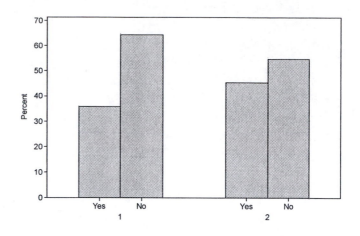

	Explanatory Variable	
Response	1	2
Yes	0.357	0.452
No	0.643	0.548
Total	1.000	1.000

9.13 (a) p_i = proportion of "yes" responses in group i. $H_0 : p_1 = p_2$ $H_a : p_1 \neq p_2$. $\hat{p}_1 = 0.357$, $\hat{p}_2 = 0.452$. **(b)** $\hat{p} = (75 + 95)/(210 + 210) = 0.4048$. $z = -1.9882$, $P = 0.0469$. We reject H_0 at the 5% level and conclude there is a difference in the proportion of "yes" answers for the two levels. **(c)** The P-values agree (software gives $P = 0.0469$ for $X^2(1) = 3.95$). **(d)** $z^2 = (-1.9882)^2 = 3.9529$.

9.15 The expected counts were rounded to the nearest hundredths place; the difference is roundoff error.

9.17 The contributions to the chi-square statistic are $\dfrac{(Obs - Exp)^2}{Exp}$. For California, this is

$\dfrac{(257 - 269.524)^2}{269.524}$. The chi-square statistic is $0.0369 + 0.5820 + 0.0000 + 0.0196 + 0.0660 + 0.2264 = 0.9309$.

State	AZ	CA	HI	IN	NV	OH
Observed count	167	257	257	297	107	482
Proportion	0.105	0.172	0.164	0.188	0.070	0.301
Expected count	164.535	269.524	256.988	294.596	109.690	471.667
Chi-square contribution	0.0369	0.5820	0.0000	0.0196	0.0660	0.2264

9.19 (a) H_0: $P(\text{Head}) = P(\text{Tail}) = 0.5$ versus H_a: H_0 is incorrect (the probabilities are not 0.5). **(b)** With $n = 10{,}000$ tosses, $E(\text{Heads}) = E(\text{Tails}) = 5000$. $X^2 = \dfrac{(5067 - 5000)^2}{5000} + \dfrac{(4933 - 5000)^2}{5000} = 1.7956$, df $= 1$. From Table F, $0.15 < P < 0.20$ (software gives $P = 0.1802$). Fail to reject H_0. We have no reason to believe the coin was not fair.

9.21 (a) See the tables below. The marginal distributions are shown as the totals row and column on the joint distribution table.

Joint Distribution	Design 1	Design 2	Total
More than 1 min.	0.24	0.10	0.34
Less than 1 min.	0.26	0.40	0.66
Total	0.50	0.50	1.00

	Design 1	Design 2
More than 1 minute	0.48	0.20
Less than 1 minute	0.52	0.80
	1.00	1.00

	Design 1	Design 2	
More than 1 min.	0.7059	0.2941	1.0000
Less than 1 min.	0.3939	0.6061	1.0000

(b) See the tables below.

Joint Distribution	1st year	4th year	Total
Yes	0.1460	0.2010	0.3471
No	0.4742	0.1787	0.6529
Total	0.6203	0.3797	1.0000

	1st year	4th year
Yes	0.2355	0.5294
No	0.7645	0.4706
	1.0000	1.0000

	1st year	4th year	Total
Yes	0.4208	0.5792	1.000
No	0.7263	0.2737	1.000

9.23 (a) There is no reason to view year of study or major as predictor and response, so we only want to explore a relationship. **(b)** There is no reason to believe that student status or fruit/vegetable consumption explains the other, so we only want to explore for any possible relationship. **(c)** It seems reasonable that time of day might explain violent content of television programs (indeed, the FCC has regulations on this). **(d)** Age would explain bad teeth (especially in that era).

9.25 The expected counts in the table below were calculated using $\dfrac{\text{(row total)(column total)}}{\text{(table total)}}$.

		Times Witnessed		
Gender	Never	Once	More than once	Total
Girls	125.503	161.725	715.773	1003
Boys	120.497	155.275	687.227	963
Total	246	317	1403	1966

9.27 (a) We test $H_0 : p_1 = p_2$ versus $H_a : p_1 \neq p_2$, where the proportions of interest are those for persons harassed in person. $\hat{p}_1 = 321/361 = 0.8892$, $\hat{p}_2 = 200/641 = 0.3120$, $\hat{p} = 521/1002 = 0.5200$. We compute $z = 17.556$, $P \approx 0$. We conclude there is an association between being harassed online and in person. It appears that those who have been harassed online are more likely to also be harassed in person. **(b)** We test $H_0 :$ There is no association between being harassed online and in person versus H_a: There is a relationship. $X^2 = 308.23$, df $= 1$, $P \approx 0$. (c) $17.556^2 = 308.21$, which agrees with X^2 to within roundoff error. (d) The count here is $n = 1002$; in Exercise 9.25, there were $n = 1003$ girls. Perhaps one girl wouldn't answer these questions.

9.29 (a) The solution to Exercise 9.27 used "harassed online" as the explanatory variable. **(b)** Changing to use "harassed in person" for the two-proportions z test gives $\hat{p}_1 = 321/521 = 0.6161$, $\hat{p}_2 = 40/481 = 0.0832$, $\hat{p} = 361/1002 = 0.3603$. We again compute $z = 17.556$, $P \approx 0$. No changes will occur in the chi-square test. **(c)** If two variables are related, the test statistic will be the same regardless of which is viewed as explanatory.

9.31 For 600 rolls, the expected count for each face would be $(1/6)(100) = 100$.

9.33 (a) One might believe that opinion depended on the type of institution. **(b)** See graph at right. Presidents at 4-year public institutions are roughly equally divided about on-line courses, with presidents at 2-year public institutions slightly in favor. Four-year private school presidents are definitely not in agreement, while those at private 2-year schools seem to think online courses are equivalent to face-to-face courses.

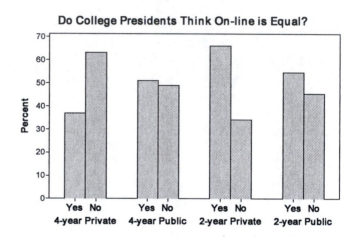

9.35 (a) Summing the "Yes" line, we find 206 presidents agreed with the question. **(b)** We have separate samples, so the two-way table is

	Presidents	Public
Yes	206	621
	(128.76)	(698.24)
No	189	1521
	(266.24)	(1443.80)

(c) The column totals for this table are the two sample sizes. The row totals might be seen as giving overall opinions on the value of on-line courses, but their use might be questionable. **(d)** We test H_0: The opinions on the value of on-line courses are the same for college presidents and the general public versus H_a: The opinions are different. Expected counts are given in the table above in parentheses. $X^2 = 81.41$, df $= 1$, $P \approx 0$. We reject H_0 and conclude there is a relationship. Looking at the observed and expected counts, college presidents are more willing to see on-line courses as equivalent than members of the general public.

9.37 (a) The 3 × 2 table is on the right. **(b)** The percents of disallowed small, medium, and large claims are (respectively) 6/57 = 10.5%, 5/17 = 29.4%, and 1/5 = 20%. **(c)** In the 3 × 2 table, the expected count for large/not allowed is too small (5)(12)/79 = 0.76. **(d)** The null hypothesis is "There is no relationship between claim size and whether a claim is allowed." **(e)** As a 2 × 2 table (with the second row 16 "yes" and 6 "no"), we find $X^2 = 3.456$, df = 1, $P = 0.063$. The evidence is not quite strong enough to reject H_0.

	Allowed?		
Stratum	Yes	No	Total
Small	51	6	57
Medium	12	5	17
Large	4	1	5
Total	67	12	79

9.39 The table on the right shows the given information translated into a 3 × 2 table. For example, in Year 1, about (0.423)(2408) = 1018.584, (which rounds to 1019) students received DFW grades, and the rest— (0.577)(2408) = 1389.416 students—passed. To test H_0: the DFW rate has not changed, we compute $X^2 = 308.3$, df = 2, $P < 0.0001$—very strong evidence of a change.

Year	DFW	Pass
1	1019	1389
2	579	1746
3	423	1703

9.41 (a) The approximate counts are shown on the right; for example, among those students in trades, (0.34)(942) = 320.28 enrolled right after high school, and (0.66)(942) = 621.72 enrolled later. **(b)** In addition to a chi-square test in part (c), students might note other things, such as overall, 39.4% of these students enrolled right after high school. Health is the most popular field, with about 38% of these students. **(c)** We have strong enough evidence to conclude that there is an association between field of study and when students enter college; the test statistic is $X^2 = 276.1$, with df = 5, for which P is very small. A graphical summary is not shown; a bar chart would be appropriate.

	Time of Entry		
Field of study	Right after high school	Later	Total
Trades	320	622	942
Design	274	310	584
Health	2034	3051	5,085
Media/IT	976	2172	3,148
Service	486	864	1,350
Other	1173	1082	2,255
Total	5263	8101	13,364

9.43 (a) The approximate counts are shown on the right; for example, among those students in trades, (0.2)(942) = 188.4 (188, after rounding) relied on parents, family, or spouse, and (0.8)(942) = 753.6 (754) did not. **(b)** We have strong enough evidence to conclude that there is an association between field of study and getting money from parents, family, or spouse; the test statistic is $X^2 = 544.0$ (with unrounded counts) or 544.8 (with rounded counts), with df = 5, for which P is very small. **(c)** Overall, 25.4% of these students relied on family support; students in media/IT and service fields were slightly less likely, and those in the design and "other" fields were slightly more likely to rely on family support. A bar graph would be a good choice for a graphical summary.

	Parents, Family, Spouse		
Field	Yes	No	Total
Trades	188	754	942
Design	222	377	599
Health	1361	3873	5,234
Media/IT	518	2720	3,238
Service	248	1130	1,378
Other	943	1357	2,300
Total	3480	10,211	13,691

9.45 (a) The percent who have lasting waking symptoms is the total of the first column divided by the grand total: $69/119 = 57.98\%$. **(b)** The percent who have both waking and bedtime symptoms is the count in the upper left divided by the grand total: $36/119 = 30.25\%$. **(c)** To test H_0: There is no relationship between waking and bedtime symptoms versus H_a: There is a relationship, we find $X^2 = 2.275$ (df = 1) and $P = 0.131$. We do not have enough evidence to conclude that there is a relationship.

Minitab Output

	WakeNo	WakeYes	All
BedNo	17	33	50
	21.01	28.99	50.00
BedYes	33	36	69
	28.99	40.01	69.00
All	50	69	119
	50.00	69.00	119.00

Cell Contents: Count
                  Expected count

Pearson Chi-Square = 2.275, DF = 1,
P-Value = 0.131

9.47 Two examples are shown on the right. In general, choose a to be any number from 0 to 100, and then all the other entries can be determined.

50	100
50	100

25	125
75	75

    Note: *This is why we say that such a table has "one degree of freedom": We can make one (nearly) arbitrary choice for the first number and then have no more decisions to make.*

9.49 The table of counts (with expected counts) is given. $X^2 = 852.433$, df = 1, $P \approx 0$. Using $z = -29.2$, computed in the solution to Exercise 8.99), this equals z^2 (up to rounding).

	Mexican American		
Selected	Yes	No	Total
Yes	339	531	870
	(688.25)	(181.75)	
No	143,272	37,393	180,665
	(142,922.75)	(37,742.25)	
Total	143,611	37,924	181,535

9.51 Answers will vary as students choose their own intervals.

9.53 Answers will vary as students generate their own samples.

9.55 (a) Each quadrant accounts for one fourth of the area, so we expect it to contain one fourth of the 100 trees. **(b)** *Some* random variation would not surprise us; we no more expect exactly 25 trees per quadrant than we would expect to see exactly 50 heads when flipping a fair coin 100 times. **(c)** The table on the right shows the individual computations, from which we obtain $X^2 = 10.8$, df = 3, and $P = 0.0129$. We conclude that the distribution is not random.

Observed	Expected	$\dfrac{(O-E)^2}{E}$
18	25	1.96
22	25	0.36
39	25	7.84
21	25	0.64
	100	10.80

Chapter 10 Solutions

10.1 (a) The slope is the coefficient of x, 3.1. **(b)** The slope of 3.1 means the *average* value of y increases 3.1 units for each unit increase in x. **(c)** $\mu_y = 51.6 + 3.1(10) = 82.6$. **(d)** 95% of the subpopulation will be between $\mu_y \pm 2\sigma$, or $82.6 \pm 2(5.2) = 72.2$ to 93.0.

10.3 (a) $t = \dfrac{1.8}{0.95} = 1.895$, df $= 25 - 2 = 23$. From Table D, we have $0.05 < P < 0.10$ (software gives 0.0707). **(b)** $t = \dfrac{2.0}{0.95} = 2.105$, df $= 25 - 2 = 23$. From Table D, $0.04 < P < 0.05$ (0.0464 from software). **(c)** $t = \dfrac{1.7}{0.55} = 3.091$, df $= 98$. Using df $= 80$ in Table D, $0.002 < P < 0.005$ (0.0026 from software).

10.5 The margin of error is half the width of the confidence interval. $(24.4 - 23.0)/2 = 0.7$ kg/m^2. The margin of error depends on the distance from $\bar{x}$. Because $x = 5.0$ is farther from the mean, the margin of error there will be larger.

10.7 (a) The scatterplot suggests a linear increase over time. **(b)** The fitted line is

$\widehat{\text{Spending}} = -4900.5333 + 2.4667\,\text{Year}$.

(**Note:** Rounding in this exercise can make a big difference in results.) **(c)** The residuals are $e_i = y_i - \hat{y}_i$.

Year	$\hat{y}$	e_i
2003	40.2	−0.1
2006	47.6	0.2
2009	55.0	−0.1

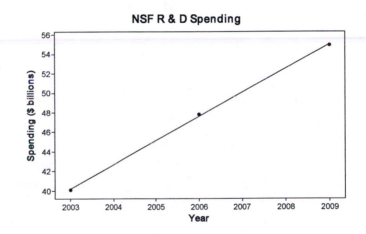

NSF R & D Spending

From these residuals, we calculate

$s = \sqrt{\dfrac{(-0.1)^2 + (0.2)^2 + (-0.1)^2}{3-2}} = 0.2449$. **(d)** The model is $y = \beta_0 + \beta_1 x + \varepsilon$. We have estimates

$\hat{\beta}_0 = -4900.5333$, $\hat{\beta}_1 = 2.4667$, and $\hat{\sigma}(\varepsilon_i) = 0.2449$. **(e)** $s(b_1) = \dfrac{s}{\sqrt{\sum(x_i - \bar{x})^2}} = \dfrac{0.2449}{\sqrt{18}} = 0.0577$.

df $= 1$, so $t^* = 12.71$. The confidence interval is $2.4667 \pm 12.71(0.0577) = 1.733$ to 3.200. This interval says that NSF research and development funding is increasing between 1.733 and 3.2 billion dollars per year, at 95% confidence.

10.9 (a) The parameters are β_0, β_1, and σ; b_0, b_1, and s are the *estimates* of those parameters. **(b)** H_0 should refer to β_1 (the population slope) rather than b_1 (the estimated slope). **(c)** The confidence interval will be narrower than the prediction interval because the confidence interval accounts only for the uncertainty in our estimate of the mean response, while the prediction interval must also account for the random error of an individual response.

10.11 Because Kiplinger narrows down the number of colleges, the selected colleges are an SRS from that list, and not from the original 500-plus 4-year public colleges.

10.13 (a) $\widehat{AvgDebt} = 11{,}818 + 168.98446(46) = \$19{,}591.29$. **(b)** $\widehat{AvgDebt} = 11{,}818 + 168.98446(69) = \$23{,}477.93$. **(c)** The center of the x distribution is roughly 50%. 69% is farther from that center than 46%, so University of Wisconsin-LaCrosse would have a larger standard error.

10.15 Prediction intervals concern individuals instead of means. Departures from the Normal distribution assumption would be more severe here (in terms of how the individuals vary around the regression line).

10.17 (a) $H_0 : \beta_1 = 0$ and $H_a : \beta_1 > 0$. It does not seem reasonable to believe that tuition will decrease. **(b)** From software, $t = 13.94$, $P < 0.0005$ (df = 26). **(c)** Using df = 26 from Table D, $0.9675 \pm 2.056(0.06939) = 0.825$ to 1.110. At 95% confidence, average tuition increased between -17.5% and 11.0% per year. (A slope less than 1 here means average tuition actually decreased.) **(d)** $r^2 = 88.2\%$. **(e)** Inference on β_0 would be extrapolation; there were no colleges close to \$0 tuition in 2008.

10.19 (a) See scatterplot. The relationship is strong (little scatter), increasing, and fairly linear; however, there may be a bit of curve at each end.

(b) $\widehat{Out11} = 1075 + 1.153\ Out08$ (or, $\hat{y} = 1075 + 1.153x$.) **(c and d)** A residuals plot and a Normal quantile plot of the residuals are shown below. No overt problems are noted, even though the Normal plot wiggles around the line. Some students may think there is a change in variability in the plot of

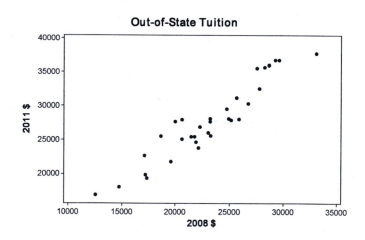

residuals against 2008 tuition, but there are fewer data points on the ends of the plot and more in the middle.

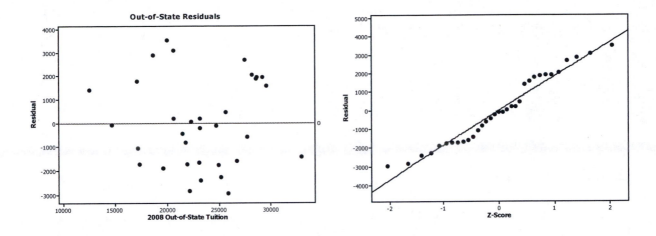

Minitab Output: Out-of-State 2011 versus Out-of-State 2008

```
The regression equation is Out11 = 1075 + 1.15 Out08

Predictor      Coef    SE Coef      T       P
Constant       1075       1700    0.63   0.532
Out08       1.15339    0.07175   16.08   0.000

S = 1888.44   R-Sq = 89.3%   R-Sq(adj) = 88.9%
```

10.21 The scatterplot shows a weak, increasing, somewhat linear relationship between in-state and out-of-state tuition rates for 2011. Minnesota appears to be an outlier with in-state tuition $13,022 and out-of-state tuition $18,022. The regression equation is $\widehat{Out11} = 17{,}160 + 1.017\, In11$ (or, $\hat{y} = 17{,}160 + 1.017x$. The scatterplot of residuals against x shows no overt problems (except the low outlier for Minnesota); the Normal quantile plot also shows no problems, although we note that several schools seem to have similar residuals (slightly more than $5000).

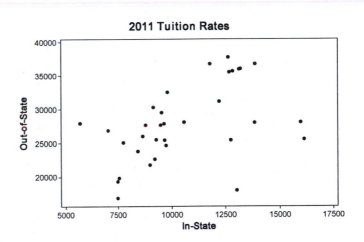

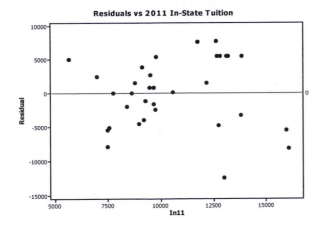

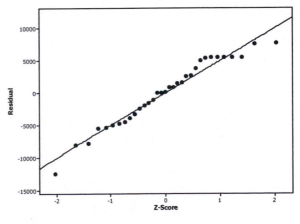

Minitab Output: Out-of-State versus In-State

```
The regression equation is Out11 = 17160 + 1.02 In11

Predictor     Coef   SE Coef     T        P
Constant     17160      3714   4.62    0.000
In11        1.0172    0.3421   2.97    0.006

S = 5089.60    R-Sq = 22.2%    R-Sq(adj) = 19.7%
```

10.23 (a) The scatterplot (with the regression line) is shown at right. There is a strong, increasing, linear relationship between BAC and beer consumption. There may be an outlier at (9, 0.19). The regression equation is $\hat{y} = -0.0127 +$ 0.0180x, and $r^2 = 80.0\%$. Not surprisingly, we find that BAC increases as beer consumption increases; the relationship is quite strong, with beer consumption explaining 80% of the variation in BAC. **(b)** To test $H_0: \beta_1 = 0$ versus $H_a: \beta_1 > 0$, we find $t = 7.48$ and

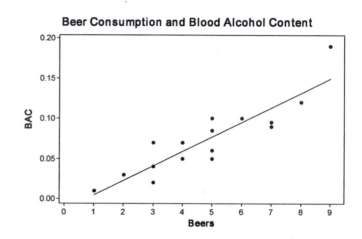

$P < 0.0001$. There is very strong evidence that drinking more beers increases BAC. **(c)** The predicted mean BAC for $x = 5$ beers is $\hat{y} = -0.0127 + 0.0180(5) = 0.07712$; the 90% prediction interval is 0.040 to 0.114. Steve might be safe, but he cannot be sure that his BAC will be below 0.08.

   Note: *We use a prediction interval (rather than a confidence interval) because we want a range of values for an* individual *BAC after 5 beers, rather than the* mean *BAC.*

Minitab Output: Regression of BAC on beer consumption

```
The regression equation is BAC = - 0.0127 + 0.0180 Beers

Predictor       Coef    SE Coef       T        P
Constant    -0.01270    0.01264   -1.00    0.332
Beers       0.017964   0.002402    7.48    0.000

S = 0.0204410    R-Sq = 80.0%    R-Sq(adj) = 78.6%

New Obs     Fit    SE Fit          90% CI                  90% PI
      1  0.07712  0.00513  (0.06808, 0.08615)  (0.04000, 0.11424)
```

10.25 (a) Histograms are provided. Both distributions are right-skewed; in fact, over half of the $n = 203$ players have less than 1.5% of their compensation based on performance bonuses. Summary statistics follow.

Variable	N	Mean	StDev	Minimum	Q1	Median	Q3	Maximum
Percentage	203	14.22	23.31	0.00	0.31	1.43	17.65	85.01
Rating	203	7.759	6.343	0.000	2.250	6.310	12.690	27.880

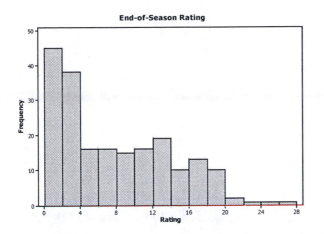

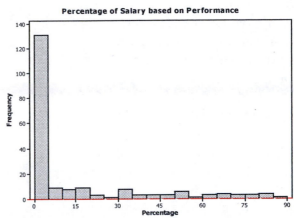

(b) The fact that these distributions are not Normal is not necessarily a problem; the condition is that the residuals are Normal. **(c)** Because we want to know if bonuses result in better performance, the performance rating is the response variable. The graph does not look like a linear relationship at all. If one exists, it is very weak. **(d)** The equation (Minitab output following) is Rating = 6.247 + 0.1063 Percentage. $r^2 = 15.3\%$ confirms the weak relationship. **(e)** Based on the histogram, the residuals are not Normally distributed, but are skewed right. A Normal quantile plot (not shown) confirms this. These data are not reasonable for regression inference.

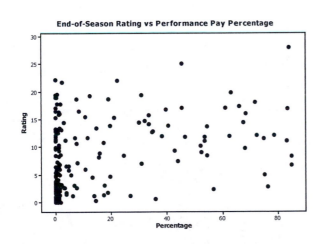

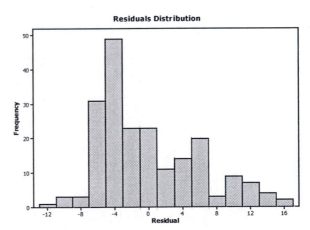

Minitab Output: Rating versus Percentage

The regression equation is Rating = 6.247 + 0.1063 Percentage

Predictor	Coef	SE Coef	T	P
Constant	6.2469	0.4816	12.97	0.000
Percentage	0.10634	0.01767	6.02	0.000

S = 5.85353   R-Sq = 15.3%   R-Sq(adj) = 14.8%

10.27 (a) 17 of the 30 (56.7%) had a selling price higher than the appraisal. This was "an SRS of 30 properties," so it should be reasonably representative. (In fact, it's typical that assessed values trail actual market values.) **(b)** The relationship is increasing, moderately strong, and linear. One property was assessed at almost $300,000 and sold for less than $200,000. **(c)** $\hat{y} = 66.95 + 0.6819x$. **(d)** The residuals from this regression are shown below, left. The outlier point is still an outlier in this plot; it is almost three standard deviations below its predicted value. **(e)** Without the outlier, the equation is $\hat{y} = 37.41 + 0.8489x$. The intercept is now just over half of what is was, and the slope has increased (if appraised value and selling price were identical, we would have a model $y = x$). The standard deviation has decreased from $s = 31.41$ to $s = 26.80$. **(f)** There are no clear violations of the assumptions—at least, none severe enough to cause too much concern.

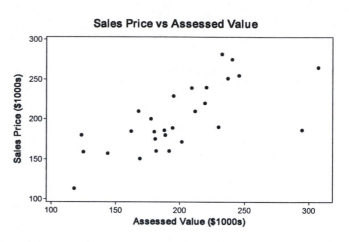

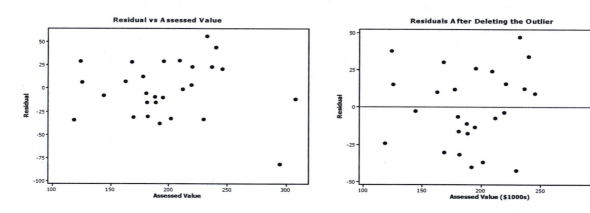

Minitab Output: All data	Minitab Output: Without the Outlier
The regression equation is Sales Price = 66.9 + 0.682 Assessed Value	The regression equation is Sales Price = 37.41 + 0.849 Assessed Value
Predictor Coef SE Coef T P Constant 66.95 26.11 2.56 0.016 Assessed 0.6819 0.1292 5.28 0.000	Predictor Coef SE Coef T P Constant 37.41 23.93 1.56 0.130 Assessed 0.8489 0.1208 7.03 0.000
S = 31.4098 R-Sq = 49.9% R-Sq(adj) = 48.1%	S = 26.8024 R-Sq = 64.7% R-Sq(adj) = 63.3%

10.29 (a) The plot could be described as moderately strong, increasing, and roughly linear, or possibly curved; it almost looks as if there are two lines; one for years before 1980 and one after that. 2012 had an unusually low number of

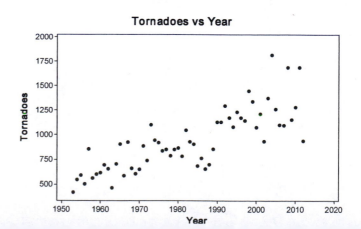

tornadoes, while 2004 had an unusually high number. **(b)** $\widehat{\text{Tornadoes}} = -27{,}432 + 14.312$ Year (or $\hat{y} = -27{,}432 + 14.312x$). The 95% confidence interval for the slope of the regression line is $14.312 \pm 2.009(1.391) = 11.52$ to 17.11 using df $= 50$ (using technology and df $= 58$, we get $14.312 \pm 2.002(1.391) = 11.53$ to 17.10). **(c)** See the plot below. We see what seems to be an increasing amount of scatter in later years; 2012 again stands out as does 2004. **(d)** Based on the Normal quantile plot, we can believe the residuals are Normally distributed; with the exception of 2004, all are very close to the line. **(e)** After eliminating 2004 and 2012 from the data set, the new equation is $\widehat{\text{Tornadoes}} = -27{,}458 + 14.324$ Year. These years are not very influential to the regression (the slope and intercept changed very little); it would seem these two years only add variability. The residuals versus year plot (not shown) looks random, and the Normal quantile plot also shows no deviations from a Normal distribution.

Minitab Output: All Years					**Regression After Removing 2004 and 2012**				
The regression equation is					The regression equation is				
Tornadoes = - 27432 + 14.3 Year					Tornadoes = - 27458 + 14.3 Year				
Predictor	Coef	SE Coef	T	P	Predictor	Coef	SE Coef	T	P
Constant	-27432	2757	-9.95	0.000	Constant	-27458	2529	-10.86	0.000
Year	14.312	1.391	10.29	0.000	Year	14.324	1.276	11.23	0.000
S = 186.54 R-Sq = 64.6% R-Sq(adj) = 64.0%					S = 164.488 R-Sq = 69.2% R-Sq(adj) = 68.7%				

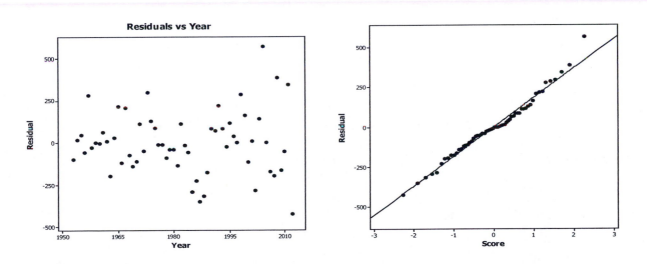

10.31 (a) About $r^2 = 8.41\%$ of the variability in AUDIT score is explained by (a linear regression on) gambling frequency. **(b)** With $r = 0.29$ and $n = 908$, the test statistic for $H_0: \rho = 0$ versus $H_a: \rho \neq 0$ is $t = r\sqrt{n-2}/\sqrt{1-r^2} = 9.12$ (df $= 906$), for which P is very small. **(c)** Even though this study found a significant positive correlation between gambling and drinking behaviors, the relationship is weak. In addition, nonresponse is a problem because the students who did not answer might have different characteristics from those who did. Because of this, we should be cautious about considering these results to be representative of all first-year students at this university, and even more cautious about extending these results to the broader population of all first-year students.

10.33 (a) The stemplot of percent forested is shown on the right; see the solution to the previous exercise for the stemplot of IBI. Percent forested (x) is right-skewed; $\bar{x} = 39.3878\%$, $s_x = 32.2043\%$. IBI is left-skewed; $\bar{y} = 65.9388$, $s_y = 18.2796$. **(b)** The scatterplot (below, left) shows a weak positive association, with more scatter in y for small x. **(c)** $y_i = \beta_0 + \beta_1 x_i + \varepsilon_i$, $i = 1, 2, ..., 49$; ε_i are independent $N(0, \sigma)$ variables. **(d)** The hypotheses are $H_0 : \beta_1 = 0$ versus $H_a : \beta_1 \neq 0$. **(e)** See the Minitab output following. We have $\widehat{\text{IBI}} = 59.91 + 0.1531$ Forest, and the estimated standard deviation is $s = 17.79$. For testing the hypotheses in (d), $t = 1.92$ and $P = 0.061$. **(f)** The residual plot (below, right) shows a slight curve—the residuals seem to be (very) slightly lower in the middle and higher on the ends. **(g)** As we can see from a stemplot and/or a Normal quantile plot (both below), the residuals are left-skewed. **(h)** Student opinions may vary. The three apparent deviations from the model are (i) a possible change in standard deviation as x changes, (ii) possible curvature of residuals, and (iii) possible non-Normality of error terms.

Percent forested

```
 0 | 00000033789
 1 | 0014778
 2 | 125
 3 | 123339
 4 | 133799
 5 | 229
 6 | 38
 7 | 599
 8 | 069
 9 | 055
10 | 00
```

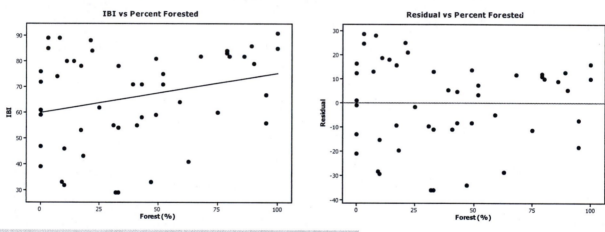

Minitab Output: Regression of IBI on Forest Percent

```
The regression equation is IBI = 59.9 + 0.153 Forest

Predictor      Coef   SE Coef       T       P
Constant     59.907     4.040   14.83   0.000
Forest      0.15313   0.07972    1.92   0.061

S = 17.7880   R-Sq = 7.3%   R-Sq(adj) = 5.3%
```

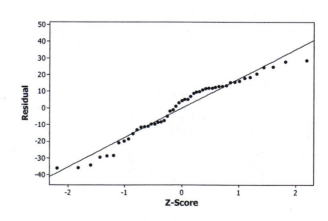

```
-3 | 55
-3 | 4
-2 | 988
-2 | 0
-1 | 985
-1 | 2110
-0 | 99887
-0 | 410
 0 | 134
 0 | 557899
 1 | 01122333
 1 | 55678
 2 | 044
 2 | 78
```

10.35 The precise results of these changes depend on which observation is changed. (There are six observations which had 0% forest and two which had 100% forest.) Specifically, if we change IBI to 0 for one of the first six observations, the resulting *P*-value is between 0.019 (observation 6) and 0.041 (observation 3). Changing one of the last two observations changes the *P*-value to 0.592 (observation 48) or 0.645 (observation 49). In general, the first change decreases *P* (that is, the relationship is more significant) because it accentuates the positive association. The second change weakens the association, so *P* increases (the relationship is less significant).

10.37 Using Area = 10 in $\widehat{IBI} = 52.92 + 0.4602$ Area from Exercise 10.32, $\widehat{IBI} = 57.52$. Using Forest = 63 in $\widehat{IBI} = 59.91 + 0.1531$ Forest from Exercise 10.33, $\widehat{IBI} = 69.55$. Both predictions have a lot of uncertainty; recall that r^2 was fairly small for both models. Also, note that the prediction intervals (shown below) are both about 70 units wide.

Minitab Output: Predicting IBI for watershed area = 10

```
   Fit   SE Fit      95% CI           95% PI
 57.52    3.41   (50.66, 64.39)   (23.56, 91.49)
```

Predicting IBI for percent forest = 63

```
   Fit   SE Fit      95% CI            95% PI
 69.55    3.16   (63.19, 75.92)   (33.21, 105.90)
```

10.39 (a) The trend appears to be quite linear. **(b)** The regression equation is $\widehat{Lean} = -61.12 + 9.3187$ Year with $s = 4.181$. The regression explains $r^2 = 98.8\%$ of the variation in lean. **(c)** The rate we seek is the slope. For df = 11 and 99% confidence, $t^* = 3.1058$, so the interval is $9.3187 \pm (3.1058)(0.3099) = 8.3562$ to 10.2812 tenths of a millimeter/year.

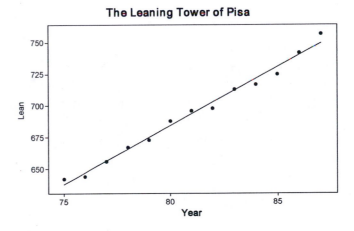

The Leaning Tower of Pisa

Minitab Output: lean versus year
```
The regression equation is lean = - 61.1 + 9.32 year

Predictor     Coef   SE Coef      T      P
Constant    -61.12    25.13   -2.43   0.033
year        9.3187   0.3099   30.07   0.000

S = 4.18097   R-Sq = 98.8%   R-Sq(adj) = 98.7%
```

10.41 (a) 2013 would be coded as 113. **(b)** The prediction is 991.89 mm beyond 2.9 m, or about 3.892 m. **(c)** We would be interested in predicting for one specific year, so we would use the prediction interval.

Minitab Output: Prediction for x = 113
```
   Fit   SE Fit        95% CI                95% PI
991.89     9.98  (969.91, 1013.87)   (968.06, 1015.72)
```

Note: *The Tower had major restoration work between 1991 and 2001 to try to halt the lean. The following was reported in August 2013 (The Guardian): "On 15 August the scientific committee charged with monitoring the tower revealed in its annual report that it had spontaneously recovered 2.5cm of its vertical incline between 2001 and 2013."*

10.43 To test $H_0 : \rho = 0$ versus $H_a : \rho \neq 0$, we compute $t = \dfrac{r\sqrt{n-2}}{\sqrt{1-r^2}} = -4.16$. Comparing this to a t distribution with df = 116, we find $P < 0.0001$, so we conclude the correlation is different from 0.

10.45 For simple linear regression, DFM = 1. Because DFT = DFM + DFE and SST = SSM + SSE, we can find the missing degrees of freedom (DF) and sum of squares (SS) entries on the Residual row by subtraction: DFE = 18 and SSE = 3304.3.

Source	DF	SS	MS	F
Model	1	4947.2	4947.2	26.95
Error	18	3304.3	183.572	
Total	19	8251.5		

The entries in the mean square (MS) column are MSM = SSM/ DFM = 4947.2 and MSE = SSE/ DFE = 183.572. Finally, F = MSM/MSE = 26.95.

10.47 As $s_x = \sqrt{1/19 \sum (x_i - \overline{x})^2} = 19.09\%$, we have $\sqrt{\sum (x_i - \overline{x})^2} = s_x \sqrt{19} = 83.2114\%$. So, $SE_{b_1} = s / \sqrt{\sum (x_i - \overline{x})^2} = 13.549/83.2114 = 0.1628$. Alternatively, note that we have $F = 26.95$ and $b_1 = 0.845$. Because $t^2 = F$, we know that $t = 5.191$ (take the positive square root, because t and b_1 have the same sign). Then, $SE_{b_1} = b_1/t = 0.1628$. (Note that with this approach, we do not need to know that $s_x = 19.09\%$.) Finally, with df = 18, $t^* = 2.101$ for 95% confidence, so the 95% confidence interval is $0.845 \pm (2.101)(0.1628) = 0.845 \pm 0.3420 = 0.503$ to 1.187.

10.49 The test statistic is $t = \dfrac{r\sqrt{n-2}}{\sqrt{1-r^2}}$. For $n = 15$, this becomes $t = 2.08$; for $n = 25$ we have

$t = 2.77$. The *P*-values are 0.0579 and 0.0109. The test with $n = 25$ is significant at the 0.05 level, but the other is not. Finding the same correlation with more data points is stronger evidence that the observed correlation is not just due to chance.

10.51 (a) The data plot is at right. There is a strong, increasing linear relationship between the two test scores. One student stands out as an outlier with an ACT score of 21 and an SAT score of 420. **(b)**

$$\widehat{ACT} = 1.63 + 0.0214 \text{ SAT (or } \hat{y} = 1.63 + 0.0214x).$$ $t = 10.78$, $P = 0.000$ (according to Minitab). ACT and SAT scores appear to be linearly related. **(c)** $r = \sqrt{0.667} = 0.8167$.

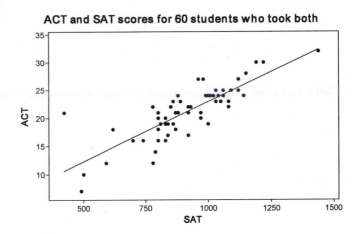

ACT and SAT scores for 60 students who took both

Minitab Output: act versus sat

```
The regression equation is act = 1.63 + 0.0214 sat

Predictor       Coef    SE Coef       T       P
Constant       1.626      1.844    0.88   0.382
sat         0.021374   0.001983   10.78   0.000

S = 2.74353    R-Sq = 66.7%    R-Sq(adj) = 66.1%
```

10.53 (a) For SAT: $\bar{x} = 912.7$ and $s_x = 180.1117$. For ACT: $\bar{y} = 21.13$ and $s_y = 4.7137$. Therefore, the slope is $a_1 = 0.02617$, and the intercept is $a_0 = -2.7522$. **(b)** The new line is dashed. **(c)** For example, the first prediction is $-2.7522 + (0.02617)(1000) = 23.42$. Up to rounding error, the mean and standard deviation of the predicted scores are the same as those of the ACT scores: $\bar{y} = 21.13$ and $s_y = 4.7137$.

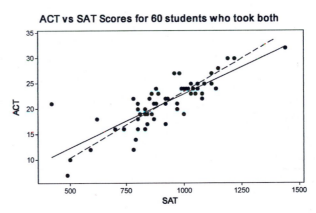

ACT vs SAT Scores for 60 students who took both

Note: *The usual least-squares line minimizes the total squared vertical distance from the points to the line. If instead we seek to minimize the total of* $\sum |h_i v_i|$, *where h_i is the horizontal distance and v_i is the vertical distance, we obtain the line $\hat{y} = a_0 + a_1 x$—except that we must choose the sign of a_1 to be the same as the sign of r. (It would hardly be the "best line" if we had a positive slope with a negative association.) If $r = 0$, either sign will do.*

10.55 (a) For squared length: $\widehat{\text{Weight}} = -117.99 + 0.4970 \text{ SQLEN}$, $s = 52.76$, $r^2 = 0.977$. **(b)** For squared width: $\widehat{\text{Weight}} = -98.99 + 18.732 \text{ SQWID}$, $s = 65.24$, $r^2 = 0.965$. Both scatterplots look more linear.

Minitab Output: weight versus Length squared	**Regression Analysis: weight versus Width squared**

```
The regression equation is
weight = - 118 + 0.497 Length squared


Predictor      Coef   SE Coef      T       P
Constant     -117.99    27.88   -4.23   0.002
Length sq    0.49701  0.02400   20.71   0.000

S = 52.755   R-Sq = 97.7   R-Sq(adj) = 97.5%
```

```
The regression equation is
weight = - 99.0 + 18.7 Width squared


Predictor      Coef   SE Coef      T       P
Constant     -98.99     33.67   -2.94   0.015
Width sq     18.732     1.126   16.64   0.000

S = 65.2381  R-Sq = 96.5%  R-Sq(adj) = 96.2%
```

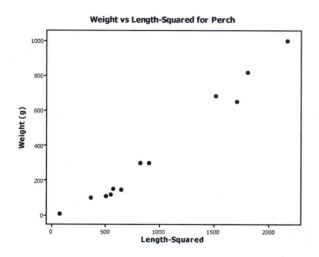

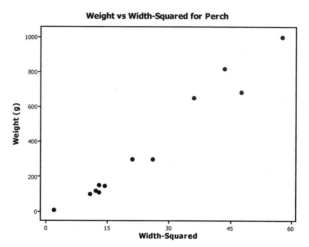

10.57 The table on the right shows the correlations and the corresponding test statistics. The first two results agree with the results of (respectively) Exercises 10.32 and 10.33.

	r	t	P
IBI/area	0.4459	3.42	0.0013
IBI/forest	0.2698	1.92	0.0608
area/forest	−0.2571	−1.82	0.0745

10.59 For each correlation, we compute

$$t = \frac{r\sqrt{n-2}}{\sqrt{1-r^2}}.$$ For the whole group, t ranges from 2.245 ($P = 0.0266$) to 3.208 ($P = 0.0017$). For Caucasians only, t ranges from 1.572 ($P = 0.1193$) to 2.397 ($P = 0.0185$). The three smallest correlations (0.16 and 0.19) are the only ones that are not significant.

Rule Breaking Measure	Popularity	Gene Expression
Sample 1 ($n = 123$)		
RB.composite	0.28**	0.26**
RB.questionnaire	0.22*	0.23*
RB.video	0.24**	0.20*
Sample 1 Caucasians only ($n = 96$)		
RB.composite	0.22*	0.23*
RB.questionnaire	0.16	0.24*
RB.video	0.19	0.16

10.61 (a) These intervals (in the table below) overlap quite a bit. **(b)** These quantities can be computed from the data, but it is somewhat simpler to recall that they can be found from the sample standard deviations $s_{x,w}$ and $s_{x,m}$:

$$s_{x,w}\sqrt{11} = 6.8684\sqrt{11} = 22.78 \text{ and } s_{x,m}\sqrt{6} = 6.6885\sqrt{6} = 16.38$$

The women's SE_{b_1} is smaller in part because it is divided by a larger *n*. **(c)** In order to reduce SE_{b_1} for men, we should choose our new sample to include men with a wider variety of lean body masses. (Note that just taking a larger sample will reduce SE_{b_1}; it is reduced even *more* if we choose subjects who will increase $s_{x,m}$.)

	b_1	SE_{b_1}	df	t^*	Interval
Women	24.026	4.174	10	2.2281	14.7257 to 33.3263
Men	16.75	10.20	5	2.5706	−9.4699 to 42.9699

Chapter 11 Solutions

11.1 (a) Second semester GPA. **(b)** $n = 242$. **(c)** $p = 7$. **(d)** Gender, standardized test score, perfectionism, self-esteem, fatigue, optimism, and depressive symptomatology.

11.3 (a) The fact that the coefficients are all positive indicates that math GPA should increase when any explanatory variable increases (as we would expect—higher values on any of these should indicate "smarter" students). **(b)** With $n = 82$ cases and $p = 4$ variables, DFM $= p = 4$ and DFE $= n - p - 1 = 77$. **(c)** In the following table, each t statistic is the estimate divided by the standard error; the P-values are computed from a t distribution with df $= 77$. (The t statistic for the intercept was not required for this exercise but is included for completeness.)

Variable	Estimate	SE	t	P
Intercept	-0.764	0.651	-1.174	0.2440
SAT Math	0.00156	0.00074	2.108	0.0383
SAT Verbal	0.00164	0.00076	2.158	0.0340
HS rank	1.470	0.430	3.419	0.0010
College placement exam	0.889	0.402	2.211	0.0300

All four coefficients are significantly different from 0 (although the intercept is not).

11.5 The correlations are summarized in the table on the right. Of the 21 possible scatterplots to be made from these seven variables, three are shown below as examples. The pairs with the largest correlations are generally easy to pick out. The whole-number scale for high school grades causes point clusters in

	HSM	HSS	HSE	SATM	SATCR	SATW
GPA	0.420	0.443	0.359	0.330	0.251	0.223
HSM		0.670	0.485	0.325	0.150	0.072
HSS			0.695	0.215	0.215	0.161
HSE				0.134	0.259	0.185
SATM					0.579	0.551
SATCR						0.734

those scatterplots and makes it difficult to determine the strength of the association. One might guess that these three scatterplots show relationships of roughly equal strength, but because of the overlapping points, the correlations are quite different; from left to right, they are 0.420, 0.443, and 0.670. We also note at least one low outlier in each plot.

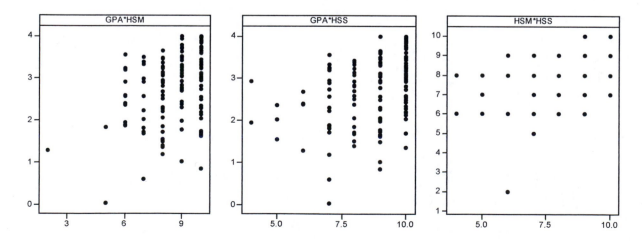

11.7 The table below gives two sets of answers: those found with critical values from Table D and those found with software. In each case, the margin of error is $t^*SE_{b_1}$, with df $= n - 3$ for parts (a) and (b), and df $= n - 4$ for parts (c) and (d). (The Table D interval for part (d) uses df $= 100$.)

	n	df	\multicolumn{2}{c}{Table D}	\multicolumn{2}{c}{Software}
			t^*	Interval	t^*	Interval
(a)	26	23	2.069	−0.0139 to 12.8139	2.0687	−0.0130 to 12.8130
(b)	53	50	2.009	0.5739 to 12.2261	2.0086	0.5751 to 12.2249
(c)	26	22	2.074	0.2372 to 9.3628	2.0739	0.2374 to 9.3626
(d)	124	120	1.984	0.6336 to 8.9664	1.9799	0.6422 to 8.9578

11.9 (a) H_0 should refer to β_2 (the population coefficient) rather than b_2 (the estimated coefficient). **(b)** This sentence should refer to the *squared* multiple correlation. **(c)** A small P implies that *at least one* coefficient is different from 0.

11.11 (a) $y_i = \beta_0 + \beta_1 x_{i1} + \beta_2 x_{i2} + \cdots + \beta_7 x_{i7} + \varepsilon_i$, where $i = 1, 2, \ldots, 142$, and ε_i are independent $N(0, \sigma)$ random variables. **(b)** The sources of variation are model (DFM $= p = 7$), error (DFE $= n - p - 1 = 134$), and total (DFT $= n - 1 = 141$).

11.13 (a) The fitted model is $\widehat{GPA} = -0.847 + 0.00269$ SATM $+ 0.229$ HSS. **(b)** $\widehat{GPA} = -0.887 + 0.00237$ SATM $+ 0.0850$ HSM $+ 0.173$ HSS. **(c)** $\widehat{GPA} = -1.11 + 0.00240$ SATM $+ 0.0827$ HSM $+ 0.133$ HSS $+ 0.0644$ HSE **(d)** $\widehat{GPA} = 0.257 + 0.125$ HSM $+ 0.172$ HSS. The "best" model has the highest predictive power (and lowest error) with the fewest predictors; this would be the model with SATM and HSS; there is not much difference between this model and the next two in terms of MSE or R^2, and it uses fewer explanatory variables. (All models were fit using Minitab.)

	MSE	R^2	$P(x_1)$	$P(x_2)$	$P(x_3)$	$P(x_4)$
(a)	0.506	25.4%	0.001	0.000		
(b)	0.501	26.6%	0.004	0.126	0.002	
(c)	0.501	27.1%	0.004	0.137	0.053	0.315
(d)	0.527	22.4%	0.024	0.003		

11.15 The first variable to leave is InAfterAid (*P*-value 0.465). Fitting the new model gives OutAfterAid (*P*-value 0.182) as the next to leave. AvgAid (*P*-value 0.184) leaves next. At that point, all variables are significant predictors. Minitab output of the final regression model is shown below.

Minitab Output: AvgDebt versus Admit, Yr4Grad, ...

```
The regression equation is
AvgDebt = - 9521 + 118 Admit + 102 Yr4Grad + 661 StudPerFac + 130 PercBorrow

Predictor     Coef  SE Coef      T      P
Constant     -9521     6712  -1.42  0.165
Admit       118.31    42.98   2.75  0.009
Yr4Grad     102.14    45.11   2.26  0.030
StudPerFac   661.4    222.2   2.98  0.005
PercBorrow  129.53    49.13   2.64  0.012

S = 3822.45   R-Sq = 42.1%   R-Sq(adj) = 35.5%
```

11.17 We have $p = 8$ explanatory variables and $n = 795$ observations. **(a)** The ANOVA F test has degrees of freedom DFM $= p = 8$ and DFE $= n - p - 1 = 786$. **(b)** This model explains only $R^2 = 7.84\%$ of the variation in energy-drink consumption; it is not very predictive. **(c)** A positive (negative) coefficient means that large values of that variable correspond to higher (lower) energy-drink consumption. Therefore, males and Hispanics consume energy drinks more frequently than females and non-Hispanics. Consumption also increases with risk-taking scores. **(d)** Within a group of students with identical (or similar) values of those other variables, energy-drink consumption increases with increasing jock identity and increasing risk taking.

11.19 We have $n = 202$, and $p = 1$ (for Model 1) or $p = 2$ (for Model 2). **(a)** For Model 1, DFE $= 200$. For Model 2, DFE $= 199$. **(b and c)** The test statistics $t = b_i / \mathrm{SE}_{b_i}$ and P-values are in the table on the right. **(d)** The relationship is still positive after adjusting for RB. When gene expression increases by 1, popularity increases by 0.204 in Model 1, and by 0.161 in Model 2 (with RB fixed).

Model	Variable	t	P
1	Gene expression	$\dfrac{0.204}{0.066} = 3.09$	0.0023
2	Gene expression	$\dfrac{0.161}{0.066} = 2.44$	0.0153
	RB.composite	$\dfrac{0.100}{0.030} = 3.33$	0.0010

11.21 (a) The regression equation is $\widehat{\mathrm{BMI}} = 23.4 - 0.682 x_1 + 0.102 x_2$, where $x_1 = (\mathrm{PA} - 8.614)$ and $x_2 = (\mathrm{PA} - 8.614)^2$. (Minitab output next page.) **(b)** The quadratic regression explains $R^2 = 17.7\%$ of the variation in BMI. **(c)** Analysis of residuals might include a stemplot, plots of residuals versus x_1 and x_2, and a Normal quantile plot. All of these appear below; none suggest any obvious causes for concern. **(d)** From the Minitab output, $t = 1.83$ with df $= 97$, for which $P = 0.070$—not significant, but close.

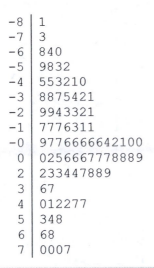

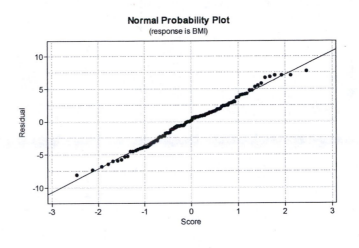

```
-8 | 1
-7 | 3
-6 | 840
-5 | 9832
-4 | 553210
-3 | 8875421
-2 | 9943321
-1 | 7776311
-0 | 9776666642100
 0 | 0256667778889
 2 | 233447889
 3 | 67
 4 | 012277
 5 | 348
 6 | 68
 7 | 0007
```

Minitab Output: BMI versus x1, x²

```
The regression equation is BMI = 23.4 - 0.682 x1 + 0.102 x^2

Predictor      Coef   SE Coef       T       P
Constant    23.3956    0.4670   50.10   0.000
X1          -0.6818    0.1572   -4.34   0.000
X^2         0.10195   0.05556    1.83   0.070

S = 3.61153   R-Sq = 17.7%   R-Sq(adj) = 16.0%
```

11.23 (a) Budget and Opening are right-skewed; Theaters and Opinion are roughly symmetric (slightly left-skewed). Five-number summaries are best, but all numerical summaries are given.

Variable	$\bar{x}$	s	Min	Q_1	M	Q_3	Max
Budget	61.81	52.47	6.5	20.0	45.0	85.0	185.0
Opening	28.59	31.89	1.1	10.0	18.6	32.1	158.4
Theaters	2785	921	808.0	2123.0	2808.0	3510.0	4366.0
Opinion	6.440	1.064	3.6	5.9	6.6	7.0	8.9

A worthwhile observation is that for all four variables, the maximum observation comes from *The Dark Knight*. **(b)** Correlations are given below. All pairs of variables are positively correlated. The Budget/Theaters and Opening/Theaters relationships appear to be curved; the others are reasonably linear.

```
    Budget                    Opening                  Theaters             Opinion
0 | 0001111          0 | 0000000011111111111     0 | 8            3 | 6
0 | 222222233        0 | 22223333                1 | 123          4 | 3
0 | 4445             0 | 455                      1 | 568          4 |
0 | 6677             0 | 666                      2 | 01444        5 | 1234
0 | 888              0 |                          2 | 556778899    5 | 5899
1 |                  1 | 0                        3 | 01244        6 | 00122
1 | 23               1 |                          3 | 55679        6 | 5566667899
1 | 44555            1 | 5                        4 | 0113         7 | 00112
1 |                                                               7 | 7
1 | 8                                                             8 | 000
```

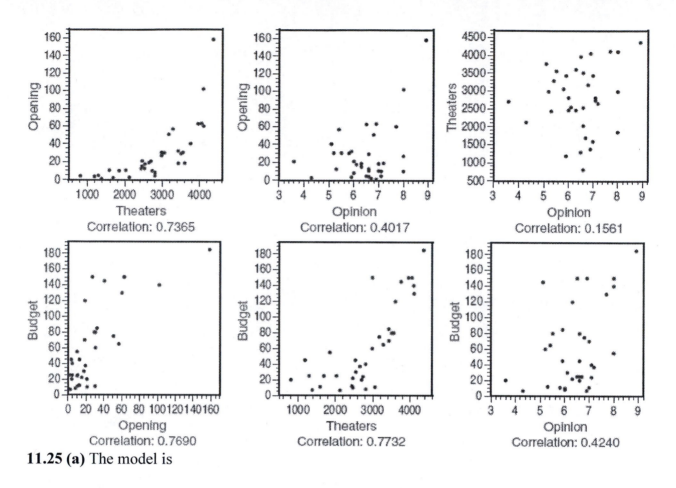

11.25 (a) The model is

$$\text{USRevenue}_i = \beta_0 + \beta_1 \text{ Budget}_i + \beta_2 \text{ Opening}_i + \beta_3 \text{ Theaters}_i + \beta_4 \text{ Opinion}_i + \varepsilon_i$$

where $i = 1, 2, \ldots, 35$, and ε_i are independent $N(0, \sigma)$ random variables. **(b)** The regression equation is

$$\widehat{\text{USRevenue}} = -67.72 + 0.1351 \text{ Budget} + 3.0165 \text{ Opening} - 0.00223 \text{ Theaters} + 10.262 \text{ Opinion}$$

(Minitab output below.) **(c)** Below is a stemplot of the residuals. The distribution is somewhat irregular, but a Normal quantile plot (not shown) does not suggest severe deviations from Normality. The residual analysis should also include a plot of residuals versus the explanatory variables; three of those plots are unremarkable (and not shown). The plot of residuals versus theaters suggests that the spread of the residuals increases with Theaters. *The Dark Knight*— noted as unusual in the previous two exercises—may be influential. **(d)** This regression explains $R^2 = 98.1\%$ of the variation in revenue.

Minitab Output: USRevenue versus Budget, Opening, Theaters, Opinion

```
The regression equation is
USRevenue = - 67.7 + 0.135 Budget + 3.02 Opening - 0.00223 Theaters + 10.3 Opinion

Predictor        Coef    SE Coef       T       P
Constant       -67.72      24.14   -2.81   0.009
Budget        0.13511    0.09776    1.38   0.177
Opening        3.0165     0.1461   20.65   0.000
Theaters    -0.002229   0.005299   -0.42   0.677
Opinion        10.262      3.032    3.38   0.002

S = 15.6929   R-Sq = 98.1%   R-Sq(adj) = 97.8%
```

```
-2 | 440
-1 | 75
-1 | 4332
-0 | 875
-0 | 3211111100
 0 | 144
 0 |
 1 | 0144
 1 | 568
 2 | 4
 2 | 8
 3 |
 3 | 6
```

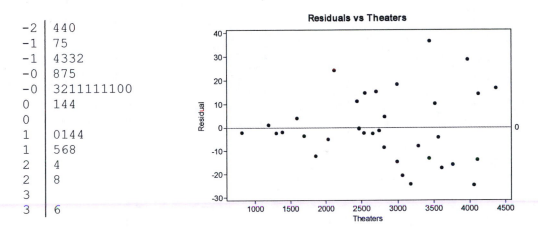

11.27 (a) Using the full model, the 95% prediction interval is $86.87 to $154.91 million. **(b)** With the reduced model, the interval is $89.94 to $154.99 million. **(c)** The intervals are very similar; as we saw in Exercise 11.26, there is little additional predictive information from the two variables we removed.

Note: *According to* http://www.imdb.com/title/tt0425061/business, *the actual U.S. revenue for* Get Smart *was $130.3 million.*

Minitab Output: Predicting U.S. revenue for *Get Smart* (full model)

```
  Fit   SE Fit          95% CI              95% PI
120.89   5.58   (109.48, 132.29)   (86.87, 154.91)
```

Minitab Output: Predicting U.S. revenue for *Get Smart* (reduced model)

```
  Fit   SE Fit          95% CI              95% PI
122.46   2.86   (116.64, 128.29)   (89.94, 154.99)
```

11.29 (a) The PEER distribution is left-skewed; the other two distributions are irregular (stemplots below). Student choices of summary statistics may vary; both five-number summaries and means/standard deviations are given below. **(b)** PEER and FtoS are slightly negatively correlated ($r = -0.114$), FtoS and CtoF are positively correlated ($r = 0.580$), and the other correlation is very small ($r = 0.004$).

Variable	$\bar{x}$	s	Min	Q_1	M	Q_3	Max
Peer review score	79.60	18.37	39	61	85	97	100
Faculty/student ratio	61.88	28.23	18	29	67	89	100
Citations/faculty ratio	63.84	25.23	17	40	66	86	100

Peer review score		Faculty/student ratio		Citations/faculty ratio	
3	9	1	8	1	79
4	2	2	1233444	2	11
4	579	2	55788889999	2	569
5	234	3	02	3	01223344
5	567777789	3	678	3	69
6	114	4		4	00124
6	57789	4	55699	4	6778
7	223	5	03	5	13
7	678	5	55779	5	89
8	023344	6		6	002223
8	566778899	6	57899	6	56667
9	0233344	7	114	7	133
9	56677889	7	5677	7	58
10	00000000000000	8	0223	8	01122333
		8	56679	8	6667999
		9	01133	9	02344
		9	5889999	9	566789
		10	000000	10	000

11.31 (a) The model is $\text{OVERALL}_i = \beta_0 + \beta_1 \text{PEER}_i + \beta_2 \text{FtoS}_i + \beta_3 \text{CtoS}_i + \varepsilon_i$, where ε_i are independent $N(0, \sigma)$ random variables. **(b)** The regression equation is

$$\widehat{\text{OVERALL}} = 18.85 + 0.5746\,\text{PEER} + 0.0013\,\text{FtoS} + 0.1369\,\text{CtoF}$$

(c) For the confidence intervals, take $b_i \pm t^* \text{SE}_{b_i}$, with $t^* = 1.994$ (for df $= 71$). These intervals have been added to the Minitab output below. The second interval contains 0 because that coefficient is not significantly different from 0. **(d)** The regression explains $R^2 = 72.2\%$ of the variation in overall score. The estimate of σ is $s = 7.043$.

Minitab Output: OVERALL versus PEER, FtoS, CtoF

```
OVERALL  =  18.8462 + 0.574625 PEER + 0.00129772 FtoS + 0.136905 CtoF
```

Term	Coef	SE Coef	T	P	95% CI	
Constant	18.8462	4.36312	4.3194	0.000		
PEER	0.5746	0.04504	12.7574	0.000	(0.4848,	0.6644)
FtoS	0.0013	0.03597	0.0361	0.971	(-0.0704,	0.0730)
CtoF	0.1369	0.03999	3.4232	0.001	(0.0572,	0.2166)

```
S = 7.04319      R-Sq = 72.22%      R-Sq(adj) = 71.04%
```

11.33 (a) All distributions are skewed to varying degrees—GINI and CORRUPT to the right, the other three to the left. CORRUPT, DEMOCRACY, and LIFE have the most skewness. Student choices of summary statistics may vary; five-number summaries are a good choice because of the skewness, but some may also give means and standard deviations.

Variable	$\bar{x}$	s	Min	Q_1	M	Q_3	Max
LSI	6.239	1.209	2.80	5.425	6.15	7.300	8.30
GINI	37.26	8.58	24.70	35.50	35.50	42.32	63.10
CORRUPT	4.910	2.451	1.70	4.15	4.150	7.175	9.60
DEMOCRACY	4.292	1.680	0.50	5.00	5.000	5.500	6.00
LIFE	72.97	8.55	47.56	70.76	74.67	79.64	82.25

Notice especially how the skewness is apparent in the five-number summaries. **(b)** There are 10 scatterplots, all of which follow. GINI is negatively (and weakly) correlated to the other four variables, while all other correlations are positive and more substantial (0.525 or more). We note a curved relationship between LIFE and CORRUPT and between DEMOCRACY and CORRUPT.

	GINI	CORRUPT	DEMOCRACY	LIFE
		Correlations		
LSI	−0.050	0.697	0.609	0.722
LIFE	−0.396	0.650	0.525	
DEMOCRACY	−0.152	0.747		
CORRUPT	−0.348			

DEMOCRACY		LIFE		GINI	
0	55	4	799	2	44
1	00	5	223	2	5556667889
1	5555555	5	99	3	00011112223333344444
2	0	6	1	3	5555556666677889
2	55	6	5667789	4	0000022344
3	000000	7	00111122222233333444	4	577788
3	5555	7	55566667778888999999	5	024
4	000	8	000000011111112	5	567
4	5555555			6	3
5	0000000				
5	5555555555555555				
6	000000000000000				

	LSI			CORRUPT
2	8		1	79
3	0		2	1111223
3	7		2	555666788899999
4	4		3	0224
4	5789		3	55577
5	0002233444		4	01122
5	55557777789999		4	56889
6	000123344		5	023
6	5677779		5	79
7	001223344		6	3
7	555667788999		6	5567
8	023		7	12333
			7	58
			8	0
			8	6778
			9	002344
			9	6

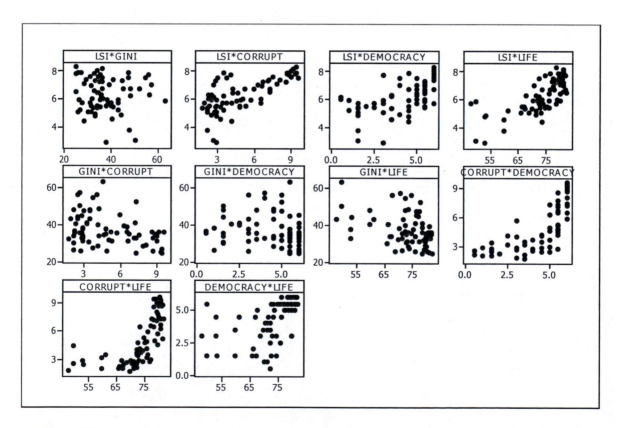

11.35 (a) The coefficients, standard errors, *t* statistics, and *P*-values are given in the Minitab output shown below. **(b)** Student observations will vary. For example, the *t* statistic for the GINI coefficient grows from $t = -0.42$ ($P = 0.675$) to $t = 4.25$ ($P < 0.0005$). The DEMOCRACY *t* is 3.53 in the third model ($P < 0.0005$) but drops to 0.71 ($P = 0.479$) in the fourth model. **(c)** A good choice is to use GINI, LIFE, and CORRUPT (Minitab output follows). All three coefficients are significant, and $R^2 = 70\%$ is nearly the same as the fourth model from Exercise 11.34. However, a scatterplot of the residuals versus CORRUPT (not shown) still looks quite a bit like the final scatterplot shown in the previous solution, suggesting a slightly curved relationship, which would violate the assumptions of our model.

Minitab Output: LSI versus GINI

```
LSI = 6.50 - 0.0071 GINI

Predictor      Coef   SE Coef      T      P
Constant     6.5028    0.6432  10.11  0.000
GINI        -0.00708   0.01683  -0.42  0.675
S = 1.21649    R-Sq = 0.3%   R-Sq(adj) = 0.0%
```

Minitab Output: LSI versus GINI, LIFE

```
LSI = - 3.82 + 0.0394 GINI + 0.118 LIFE

Predictor      Coef   SE Coef      T      P
Constant     -3.820    1.125   -3.40  0.001
GINI         0.03941   0.01188   3.32  0.001
LIFE         0.11772   0.01191   9.88  0.000
S = 0.788438   R-Sq = 58.7%   R-Sq(adj) = 57.5%
```

Minitab Output: LSI versus GINI, LIFE, DEMOCRACY

```
LSI = - 2.94 + 0.0366 GINI + 0.0945 LIFE + 0.215 DEMOCRACY

Predictor      Coef   SE Coef      T      P
Constant     -2.944    1.071   -2.75  0.008
GINI         0.03664   0.01103   3.32  0.001
LIFE         0.09451   0.01284   7.36  0.000
DEMOCRACY    0.21458   0.06074   3.53  0.001
S = 0.730038   R-Sq = 65.1%   R-Sq(adj) = 63.6%
```

Minitab Output: LSI versus GINI, LIFE, DEMOCRACY, CORRUPT

```
LSI = - 2.31 + 0.0447 GINI + 0.0782 LIFE + 0.0526 DEMOCRACY + 0.194 CORRUPT

Predictor      Coef   SE Coef      T      P
Constant     -2.314    1.013   -2.28  0.026
GINI         0.04471   0.01053   4.25  0.000
LIFE         0.07822   0.01287   6.08  0.000
DEMOCRACY    0.05261   0.07391   0.71  0.479
CORRUPT      0.19414   0.05710   3.40  0.001
S = 0.679202   R-Sq = 70.2%   R-Sq(adj) = 68.5%
```

Minitab Output: LSI versus GINI, LIFE, CORRUPT

```
The regression equation is LSI = - 2.35 + 0.0462 GINI + 0.0793 LIFE + 0.220 CORRUPT

Predictor      Coef   SE Coef      T      P
Constant     -2.355    1.008   -2.34  0.022
GINI         0.04619   0.01028   4.49  0.000
LIFE         0.07935   0.01273   6.24  0.000
CORRUPT      0.22034   0.04350   5.07  0.000

S = 0.676734   R-Sq = 70.0%   R-Sq(adj) = 68.7%
```

11.37 (a) The scatterplot in Exercise 11.36 suggests greater variation in VO+ for large OC. The regression equation is $\widehat{\text{VO}+} = 334.0 + 19.505\ \text{OC}$ with $s = 443.3$ and $R^2 = 0.435$; the test statistic for the slope is $t = 4.73$ ($P < 0.0005$), so we conclude the slope is not zero. The

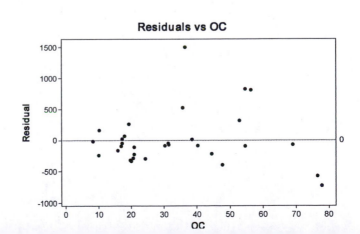

plot of residuals against OC suggests a slight downward curve on the right end, as well as wider scatter as in the middle of the OC range. The residuals are also somewhat right-skewed. A stemplot and Normal quantile plot of the residuals are not shown here but could be included as part of the analysis. **(b)** The regression equation is $\widehat{VO+} = 57.7 + 6.415\ OC + 53.87\ TRAP$ with $s = 376.3$ and $R^2 = 0.607$. The coefficient of OC is not significantly different from 0 ($t = 1.25$, $P = 0.221$), but the coefficient of TRAP is ($t = 3.50$, $P = 0.002$). This is consistent with the correlations found in the solution to Exercise 11.36: TRAP is more highly correlated with VO+, and it is also highly correlated with OC, so it is reasonable that, if TRAP is present in the model, little additional information is gained from OC (this is a case of multicollinearity among the predictors).

Minitab Output: voplus versus oc

```
voplus = 334 + 19.5 oc

Predictor     Coef   SE Coef     T       P
Constant     334.0     159.2   2.10   0.045
oc           19.505    4.127   4.73   0.000


S = 443.274    R-Sq = 43.5%    R-Sq(adj) = 41.6%
```

Minitab Output: voplus versus oc, trap

```
voplus = 58 + 6.41 oc + 53.9 trap

Predictor     Coef   SE Coef     T       P
Constant      57.7     156.5   0.37   0.715
oc            6.415     5.125   1.25   0.221
trap         53.87     15.39   3.50   0.002


S = 376.265    R-Sq = 60.7%    R-Sq(adj) = 57.9%
```

11.39 Stemplots (below) show that all four variables are noticeably less skewed.

Variable	$\bar{x}$	s	Min	Q_1	M	Q_3	Max
LVO+	6.7418	0.5555	5.652	6.240	6.768	7.132	7.842
LVO–	6.6816	0.4832	5.537	6.284	6.806	6.935	7.712
LOC	3.3380	0.6085	2.092	2.885	3.408	3.865	4.355
LTRAP	2.4674	0.4978	1.194	2.175	2.332	2.944	3.360

Scatterplots and correlations (below) show that all six pairs of variables are positively associated. The strongest association is between LVO+ and LVO–, and the weakest is between LOC and LVO–. The regression equations for these transformed variables are given in the table below, along with significance test results. Residual analysis for these regressions is not shown.

The final conclusion is the same as for the untransformed data: When we use all three explanatory variables to predict LVO+, the coefficient of LTRAP is not significantly different from 0; we then find that the model that uses LOC and LVO– to predict LVO+ is nearly as good (in terms of R^2), making it the best of the bunch.

```
    LVO+          LVO–           LOC          LTRAP
  5 | 6         5 | 5          2 | 0         1 | 1
  5 | 99        5 |            2 | 23        1 |
  6 | 011       5 | 8          2 |           1 |
  6 | 223       6 | 001        2 | 7         1 | 7
```

6	4455		6	2223		2	8888999		1	89
6	67777		6	455		3	001		2	001111
6	889		6	677		3			2	22233333
7	0011		6	8888888999		3	44455		2	
7	33		7	01		3	667		2	6667
7	4		7	23		3	89		2	99999
7	77		7	4		4	000		3	1
7	8		7	7		4	233		3	223

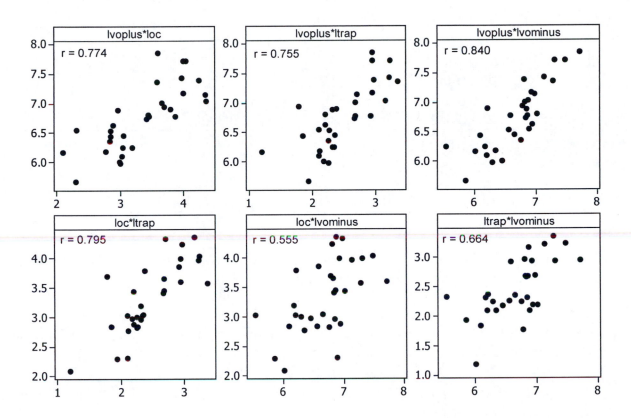

	R^2	s
$\widehat{\text{LVO}+} = 4.3841 + 0.7063\ \text{LOC}$	0.599	0.3580
$\qquad\qquad\quad$ SE = 0.1074		
$\qquad\qquad\quad$ t = 6.58		
$\qquad\qquad\quad$ P < 0.0005		

	R^2	s
$\widehat{\text{LVO}+} = 4.2590 + 0.4304\ \text{LOC}\qquad + 0.4240\ \text{LTRAP}$	0.652	0.3394
$\qquad\qquad\quad$ SE = 0.1680 $\qquad$ SE = 0.2054		
$\qquad\qquad\quad$ t = 2.56 $\qquad\qquad$ t = 2.06		
$\qquad\qquad\quad$ P = 0.016 $\qquad\qquad$ P = 0.048		

	R^2	s
$\widehat{\text{LVO}+} = 0.8716 + 0.3922\ \text{LOC}\ + 0.0275\ \text{LTRAP}\ + 0.6725$ LVOminus	0.842	0.2326
$\qquad\qquad\quad$ SE = 0.1154 $\quad$ SE = 0.1570 $\quad$ SE = 0.1178		
$\qquad\qquad\quad$ t = 3.40 $\qquad\quad$ t = 0.18 $\qquad\quad$ t = 5.71		
$\qquad\qquad\quad$ P = 0.002 $\qquad\quad$ P = 0.862 $\qquad$ P < 0.0005		

	R^2	s
$\widehat{\text{LVO}+} = 0.8321 + 0.4061\ \text{LOC}\qquad\qquad\qquad + 0.6816$ LVOminus	0.842	0.2286
$\qquad\qquad\quad$ SE = 0.0824 $\qquad\qquad\qquad$ SE = 0.1038		
$\qquad\qquad\quad$ t = 4.93 $\qquad\qquad\qquad\quad$ t = 6.57		
$\qquad\qquad\quad$ P < 0.0005 $\qquad\qquad\quad$ P < 0.0005		

11.41 Refer to the solution to Exercise 11.39 for the scatterplots. As in Exercise 11.40, the most logical single-variable model would be to use LTRAP to predict LVO–, but many students might miss that detail. Both single-explanatory variable models are given in the second table that follows. Residual analysis plots are not included. This time, we might conclude that the best model is to predict LVO– from LVO+ alone; neither biomarker variable makes an indispensable contribution to the prediction.

	R^2	s
$\widehat{\text{LVO}-} = 5.2110 + 0.4406\ \text{LOC}$	0.308	0.4089
$\qquad\qquad\quad$ SE = 0.1227		
$\qquad\qquad\quad$ t = 3.59		
$\qquad\qquad\quad$ P = 0.001		

	R^2	s
$\widehat{\text{LVO}-} = 5.0905 \qquad\qquad\qquad + 0.6449\ \text{LTRAP}$	0.441	0.3674
$\qquad\qquad\qquad\qquad\qquad\qquad$ SE = 0.1347		
$\qquad\qquad\qquad\qquad\qquad\qquad$ t = 4.79		
$\qquad\qquad\qquad\qquad\qquad\qquad$ P < 0.0005		

	R^2	s
$\widehat{\text{LVO}-} = 5.0370 + 0.0569\ \text{LOC}\qquad + 0.5896\ \text{LTRAP}$	0.443	0.3732
$\qquad\qquad\quad$ SE = 0.1848 $\qquad$ SE = 0.2259		
$\qquad\qquad\quad$ t = 0.31 $\qquad\qquad$ t = 2.61		
$\qquad\qquad\quad$ P = 0.761 $\qquad\qquad$ P = 0.014		

	R^2	s
$\widehat{\text{LVO}-} = 1.5729 - 0.2932\ \text{LOC}\ + 0.2447\ \text{LTRAP}\ + 0.8134\ \text{LVO}+$	0.748	0.2558
$\qquad\qquad\quad$ SE = 0.1407 $\quad$ SE = 0.1662 $\quad$ SE = 0.1425		
$\qquad\qquad\quad$ t = −2.08 $\qquad\quad$ t = 1.47 $\qquad\quad$ t = 5.71		

$P = 0.047$	$P = 0.152$	$P < 0.0005$	0.728	0.2611
$\widehat{\text{LVO}-} = 1.3109 - 0.1878\ \text{LOC}$		$+\ 0.8896\ \text{LVO}+$		
	$\text{SE} = 0.1237$	$\text{SE} = 0.1355$		
	$t = -1.52$	$t = 6.57$		
	$P = 0.140$	$P < 0.0005$		
$\widehat{\text{LVO}-} = 1.7570$		$+\ 0.7304\ \text{LVO}+$	0.705	0.2669
		$\text{SE} = 0.0877$		
		$t = 8.33$		
		$P < 0.0005$		

11.43 (a) The model is: $\text{PCB}_i = \beta_0 + \beta_1\ \text{PCB52}_i + \beta_2\ \text{PCB118}_i + \beta_3\ \text{PCB138}_i + \beta_4\ \text{PCB180}_i + \varepsilon_i$ where $i = 1, 2, \ldots, 69$; ε_i are independent $N(0, \sigma)$ random variables. **(b)** The regression equation is:

$\widehat{\text{PCB}} = 0.937 + 11.8727\ \text{PCB52} + 3.7611\ \text{PCB118} + 3.8842\ \text{PCB138} + 4.1823\ \text{PCB180}$ with $s = 6.382$ and $R^2 = 0.989$. All coefficients are significantly different from 0, although the constant 0.937 is not

```
-2 | 2
-1 |
-1 | 31
-0 | 8776655
-0 | 4443333222211111111000000
 0 | 0000000000001111222223333444
 0 | 677778
 1 | 12
 1 |
 2 | 2
```

$(t = 0.76, P = 0.449)$. That makes some sense—if none of these four congeners are present, it might be somewhat reasonable to predict that the total amount of PCB is 0. **(c)** The residuals appear to be roughly Normal, but with two outliers. There are no clear patterns when plotted against the explanatory variables (these plots are not shown).

Minitab Output: pcb versus pcb52, pcb118, pcb138, pcb180

```
The regression equation is
pcb = 0.94 + 11.9 pcb52 + 3.76 pcb118 + 3.88 pcb138 + 4.18 pcb180

Predictor     Coef   SE Coef       T       P
Constant     0.937     1.229    0.76   0.449
pcb52      11.8727    0.7290   16.29   0.000
pcb118      3.7611    0.6424    5.85   0.000
pcb138      3.8842    0.4978    7.80   0.000
pcb180      4.1823    0.4318    9.69   0.000

S = 6.38208   R-Sq = 98.9%   R-Sq(adj) = 98.8%
```

11.45 (a) The regression equation is

$$\widehat{\text{PCB}} = -1.018 + 12.644\ \text{PCB52} + 0.3131\ \text{PCB118} + 8.2546\ \text{PCB138}$$

with $s = 9.945$ and $R^2 = 0.973$. Residual analysis (not shown) suggests a few areas of concern: The distribution of residuals has heavier tails than a Normal distribution, and the scatter (that is, prediction error) is greater for larger values of the predicted PCB. **(b)** The estimated coefficient of PCB118 is $= 0.3131$; its P-value is 0.708. (Details in Minitab output.) **(c)** In Exercise 11.43,

$b_2 = 3.7611$ and $P < 0.0005$. **(d)** This illustrates how complicated multiple regression can be: When we add PCB180 to the model, it complements PCB118, making it useful for prediction.

Minitab Output: pcb versus pcb52, pcb118, pcb138

```
pcb = - 1.02 + 12.6 pcb52 + 0.313 pcb118 + 8.25 pcb138

Predictor    Coef   SE Coef      T      P
Constant   -1.018     1.890  -0.54  0.592
pcb52      12.644     1.129  11.20  0.000
pcb118     0.3131    0.8333   0.38  0.708
pcb138     8.2546    0.3279  25.18  0.000

S = 9.94500    R-Sq = 97.3%    R-Sq(adj) = 97.2%
```

11.47 The model is: $TEQ_i = \beta_0 + \beta_1 PCB52 + \beta_2 PCB118$ $+ \beta_3 PCB138 + \beta_4 PCB180 + \varepsilon_i$ where $i = 1, 2, \ldots, 69$; ε_i are independent $N(0, \sigma)$ random variables. The regression equation is: $\widehat{TEQ} = 1.06 - 0.0973 PCB52 + 0.3062 PCB118 + 0.1058 PCB138 - 0.0039 PCB180$ with $s = 0.9576$ and $R^2 = 0.677$. Only the constant and the PCB118 coefficient are significantly different from 0; see Minitab output below. Residuals (stemplot on the right) are slightly right-skewed and show no clear patterns when plotted with the explanatory variables (not shown).

```
-1 | 66
-1 | 42000
-0 | 9876666666666555555
-0 | 44444333221111100
 0 | 0000222224
 0 | 566667788
 1 | 23334
 1 | 9
 2 | 3
 2 | 57
```

Minitab Output: teq versus pcb52, pcb118, pcb138, pcb180

```
teq = 1.06 - 0.097 pcb52 + 0.306 pcb118 + 0.106 pcb138 - 0.0039 pcb180

Predictor      Coef   SE Coef      T      P
Constant     1.0600    0.1845   5.75  0.000
pcb52       -0.0973    0.1094  -0.89  0.377
pcb118      0.30618   0.09639   3.18  0.002
pcb138      0.10579   0.07470   1.42  0.162
pcb180     -0.00391   0.06478  -0.06  0.952

S = 0.957624    R-Sq = 67.7%    R-Sq(adj) = 65.7%
```

11.49 (a) The correlations (all positive) are listed in the table below. The largest correlation is 0.956 (LPCB and LPCB138); the smallest (0.227, for LPCB28 and LPCB180) is not quite significantly different from 0 ($t = 1.91$, $P = 0.0607$) but, with 28 correlations, such a P-value could easily arise by chance, so we would not necessarily conclude that $\rho = 0$ (or that $\rho \neq 0$).

Rather than showing all 28 scatterplots—which are all fairly linear and confirm the positive associations suggested by the correlations—we have included only two of the interesting ones: LPCB against LPCB28 and LPCB against LPCB126. The former is notable because of one outlier (specimen 39) in LPCB28; the latter stands out because of the "stack" of values in the LPCB126 data set that arose from the adjustment of the zero terms. (The outlier in LPCB28 and the stack in LPCB126 can be seen in other plots involving those variables; the two plots shown are the most appropriate for using the PCB congeners to predict LPCB, as the next exercise asks.) **(b)** All correlations are higher with the transformed data. In part, this is because these scatterplots do not exhibit the "greater scatter in the upper right" that was seen in many of the scatterplots of the original data.

	LPCB28	LPCB52	LPCB118	LPCB126	LPCB138	LPCB153	LPCB180
LPCB52	0.795						
LPCB118	0.533	0.671					
LPCB126	0.272	0.331	0.739				
LPCB138	0.387	0.540	0.890	0.792			
LPCB153	0.326	0.519	0.780	0.647	0.922		
LPCB180	0.227	0.301	0.654	0.695	0.896	0.867	
LPCB	0.570	0.701	0.906	0.729	0.956	0.905	0.829

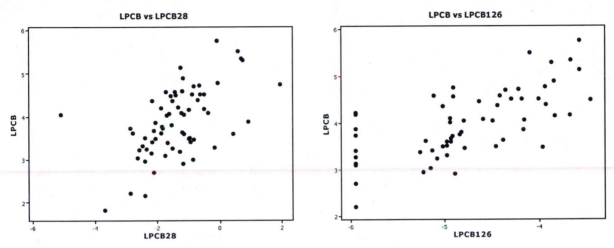

11.51 Using Minitab's BREG (best subsets regression) command for guidance, we see that there is little improvement in R^2 beyond models with four explanatory variables. The best model for each number of variables is given in the Minitab output below.

Minitab Output: Best Subsets Regression

```
                                      L L L L L
                                  L L P P P P P
                                  P P C B C B C
                                  C C B C B C B
                                  B B 1 1 1 1 1
                         Mallows  2 5 1 2 3 5 8
Vars  R-Sq  R-Sq(adj)     Cp       S  8 2 8 6 8 3 0
   1  72.9     72.5     10.8  0.31267        X
   2  76.8     76.1      1.9  0.29164  X     X
   3  77.6     76.6      1.6  0.28858  X   X X
   4  78.0     76.7      2.5  0.28815  X   X X   X
   5  78.1     76.4      4.2  0.28981  X X X X   X
   6  78.2     76.1      6.1  0.29187  X X X X   X X
   7  78.2     75.7      8.0  0.29400  X X X X X X X
```

Minitab Output: LTEQ versus LPCB28, LPBC126

```
LTEQ = 3.96 + 0.108 LPCB28 + 0.622 LPBC126

Predictor     Coef   SE Coef       T       P
Constant    3.9635    0.2275   17.42   0.000
LPCB28      0.10770   0.03246    3.32   0.001
LPBC126     0.62222   0.04801   12.96   0.000
```

```
S = 0.291639    R-Sq = 76.8%    R-Sq(adj) = 76.1%
```

Minitab Output: LTEQ versus LPCB28, LPCB118, LPBC126

```
LTEQ = 3.45 + 0.0779 LPCB28 + 0.114 LPCB118 + 0.543 LPBC126

Predictor      Coef   SE Coef       T       P
Constant     3.4450    0.4030    8.55   0.000
LPCB28      0.07792   0.03742    2.08   0.041
LPCB118     0.11356   0.07321    1.55   0.126
LPBC126     0.54348   0.06952    7.82   0.000

S = 0.288581    R-Sq = 77.6%    R-Sq(adj) = 76.6%
```

Minitab Output: LTEQ versus LPCB28, LPCB118, LPBC126, LPBC153

```
LTEQ = 3.56 + 0.0721 LPCB28 + 0.170 LPCB118 + 0.554 LPBC126 - 0.0694 LPBC153

Predictor      Coef   SE Coef       T       P
Constant     3.5573    0.4153    8.57   0.000
LPCB28      0.07215   0.03773    1.91   0.060
LPCB118     0.16967   0.08931    1.90   0.062
LPBC126     0.55378   0.07005    7.91   0.000
LPBC153    -0.06936   0.06343   -1.09   0.278

S = 0.288147    R-Sq = 78.0%    R-Sq(adj) = 76.7%
```

11.53 In the table, two *IQR*s are given; those in parentheses are based on quartiles reported by Minitab, which computes quartiles in a slightly different way from this text's method. None of the variables show striking deviations from Normality

	$\bar{x}$	M	s	IQR
Taste	24.53	20.95	16.26	23.9 (or 24.57)
Acetic	5.498	5.425	0.571	0.656 (or 0.713)
H2S	5.942	5.329	2.127	3.689 (or 3.766)
Lactic	1.442	1.450	0.3035	0.430 (or 0.4625)

in the quantile plots (not shown). Taste and H2S are slightly right-skewed, and Acetic has an irregular shape. There are no outliers.

Taste		Acetic		H2S		Lactic	
0	556	4	455	2	9	8	6
1	1234	4	67	3	1268899	9	9
1	55688	4	8	4	7799	10	689
2	011	5	1	5	024	11	56
2	556	5	2222333	6	11679	12	5599
3	24	5	444	7	4699	13	013
3	789	5	677	8	7	14	469
4	0	5	888	9	025	15	2378
4	7	6	0011	10	1	16	38
5	4	6	3			17	248
5	67	6	44			18	1
						19	09
						20	1

11.55 The regression equation is $\widehat{\text{Taste}} = -61.50 + 15.648$ Acetic with $s = 13.82$ and $R^2 = 0.302$. The slope is significantly different from 0 ($t = 3.48$, $P = 0.002$).

Based on a stemplot (right) and quantile plot (not shown), the residuals seem to have a Normal distribution. Scatterplots (below) reveal positive associations between residuals and both H2S and Lactic. The plot of residuals against Acetic suggests greater scatter in the residuals for large Acetic values.

```
-2 | 9
-2 | 11
-1 | 65
-1 | 31
-0 | 7655
-0 | 21
 0 | 012222
 0 | 5668
 1 |
 1 | 5679
 2 | 0
 2 | 6
```

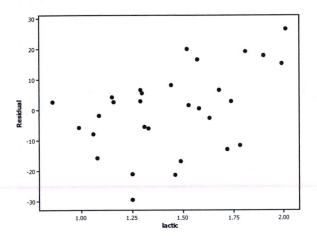

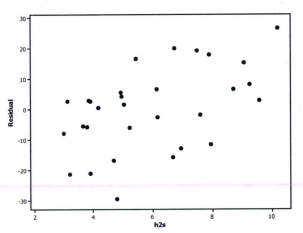

11.57 The regression equation is $\widehat{\text{Taste}} = -29.86 + 37.720$ Lactic with $s = 11.75$ and $R^2 = 0.496$. The slope is significantly different from 0 ($t = 5.25$, $P < 0.0005$).

Based on a stemplot (right) and quantile plot (not shown), the residuals appear to be roughly Normal. Scatterplots (below) reveal no striking patterns for residuals versus Acetic and H2S.

```
-1 | 965
-1 | 331
-0 | 988665
-0 | 210
 0 | 0122
 0 | 567999
 1 | 04
 1 | 58
 2 |
 2 | 7
```

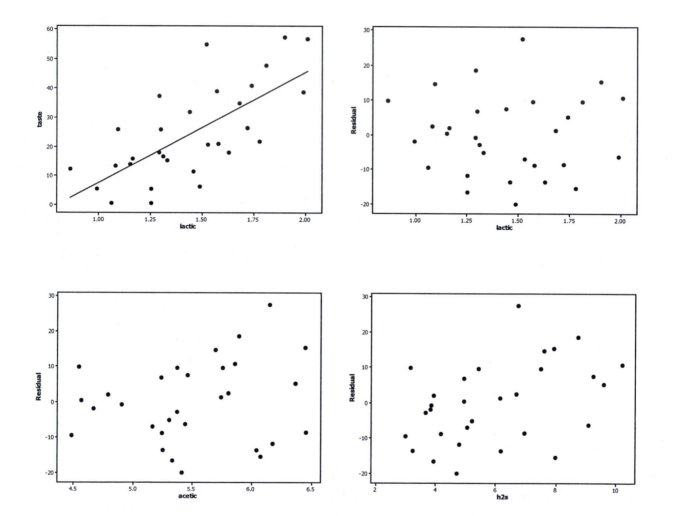

11.59 The regression equation is $\widehat{\text{Taste}} = -26.94 + 3.801$ Acetic $+ 5.146$ H2S with $s = 10.89$ and $R^2 = 0.582$. The t value for the coefficient of Acetic is 0.84 ($P = 0.406$), indicating that it does not add significantly to the model when H2S is used because Acetic and H2S are correlated (in fact, $r = 0.618$ for these two variables). This model does a better job than any of the three simple linear regression models, but it is not much better than the model with H2S alone (which explained 57.1% of the variation in Taste)—as we might expect from the t test result.

Minitab Output: Regression of taste on acetic and h2s
```
The regression equation is taste = - 26.9 + 3.80 acetic + 5.15 h2s

Predictor     Coef  SE Coef      T      P
Constant    -26.94    21.19  -1.27  0.215
acetic       3.801    4.505   0.84  0.406
h2s          5.146    1.209   4.26  0.000

S = 10.8896   R-Sq = 58.2%   R-Sq(adj) = 55.1%
```

11.61 The regression equation is $\overset{\frown}{\text{Taste}} = -28.88 + 0.328$ Acetic $+ 3.912$ H2S $+ 19.671$ Lactic with $s = 10.13$. The model explains 65.2% of the variation in Taste (the same as for the model with only H2S and Lactic). Residuals of this regression appear to be Normally distributed and show no patterns in scatterplots with the explanatory variables. (These plots are not shown.)

The coefficient of Acetic is not significantly different from 0 ($P = 0.942$); there is no gain in adding Acetic to the model with H2S and Lactic. It appears that the best model is the H2S/Lactic model of Exercise 11.60.

Minitab Output: Regression of taste on acetic, h2s, lactic
```
The regression equation is
taste = - 28.9 + 0.33 acetic + 3.91 h2s + 19.7 lactic

Predictor     Coef  SE Coef      T      P
Constant    -28.88    19.74  -1.46  0.155
acetic       0.328    4.460   0.07  0.942
h2s          3.912    1.248   3.13  0.004
lactic      19.671    8.629   2.28  0.031

S = 10.1307   R-Sq = 65.2%   R-Sq(adj) = 61.2%
```

Chapter 12 Solutions

12.1 (a) H_0 says the *population* means are all equal. **(b)** *Experiments* are best for establishing causation. **(c)** ANOVA is used to compare *means* (and assumes that the variances are equal). **(d)** Multiple comparisons procedures are used when we wish to determine which means are significantly different, but we have no specific relations in mind before looking at the data. (Contrasts are used when we have prior expectations about the differences. See page 663 for details.)

12.3 We were given sample sizes $n_1 = 23$, $n_2 = 21$, and $n_3 = 27$ and standard deviations $s_1 = 4$, $s_2 = 5$, and $s_3 = 7$. **(a)** Yes. the guidelines for pooling standard deviations say that the ratio of largest to smallest should be less than 2; $7/4 = 1.75 < 2$. **(b)** Squaring the three standard deviations gives $s_1^2 = 16$, $s_2^2 = 25$, and $s_3^2 = 49$. **(c)** $s_p^2 = \dfrac{(23-1)16 + (21-1)25 + (27-1)49}{23 + 21 + 27 - 3} =$ 31.2647. **(d)** $s_p = \sqrt{s_p^2} = 5.5915$.

12.5 (a) This sentence describes *between*-group variation. Within-group variation is the variation that occurs by chance among members of the same group. **(b)** The *sums of* squares (not the mean squares) in an ANOVA table will add; that is, SST = SSG + SSE. **(c)** The common population standard deviation σ (not its estimate s_p) is a parameter. **(d)** A small P means the means are not all the same, but the distributions may still overlap quite a bit. (See the "Caution" immediately preceding this exercise in the text.)

12.7 Assuming the t (ANOVA) test establishes that the means are different, contrasts and multiple comparison provide no further useful information. (With two means, there is only one comparison to make, and it has already been made by the t test.).

12.9 (a) With $I = 4$ groups and $N = 24$, we have df $I - 1 = 3$ and $N - I = 20$. In Table E, we see that $3.10 < F < 3.86$. **(b)** The sketch on the right shows the observed F value and the critical values from Table E. **(c)** $0.025 < P < 0.050$ (software gives 0.0463). **(d)** The alternative hypothesis states that at least one mean is different, not that all means are different.

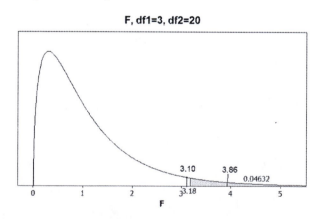

F, df1=3, df2=20

12.11 (a) $I = 4$ and $N = 64$, so the degrees of freedom are 3 and 60. $F = 127/50 = 2.54$. Comparing to the $F(3, 60)$ distribution in Table E, we find $2.18 < F < 2.76$, so $0.050 < P < 0.100$. (Software gives $P = 0.0649$.) **(b)** $I = 3$ and $N = 27$, so the degrees of freedom are 2 and 24. $F = \dfrac{58/2}{172/24} = 4.047$. Comparing to the $F(2, 24)$ distribution in Table E, we find $3.40 < F < 4.32$, so $0.025 < P < 0.050$. (Software gives $P = 0.0306$.)

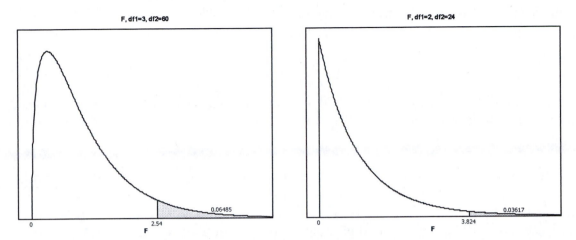

12.13 (a) Response: egg cholesterol level. Populations: chickens with different diets or drugs. $I = 3$, $n_1 = n_2 = n_3 = 25$, $N = 75$. **(b)** Response: rating on seven-point scale. Populations: the three groups of students. $I = 3$, $n_1 = 31$, $n_2 = 18$, $n_3 = 45$, $N = 94$. **(c)** Response: quiz score. Populations: students in each TA group. $I = 3$, $n_1 = n_2 = n_3 = 14$, $N = 42$.

12.15 For all three situations, the hypotheses are $H_0 : \mu_1 = \mu_2 = \mu_3$ versus H_a: at least one mean is different. The degrees of freedom are DFG = DFM = $I - 1 = 2$ ("model" or "between groups"), DFE = DFW = $N - I$ ("error" or "within groups"), and DFT = $N - 1$ ("total"). The degrees of freedom for the F test are DFG and DFE.

Situation	I	N	DFG	DFE	DFT	df for F statistic
(a) Egg cholesterol level	3	75	2	72	74	$F(2, 72)$
(b) Student opinions	3	94	2	91	93	$F(2, 91)$
(c) Teaching assistants	3	42	2	39	41	$F(2, 39)$

12.17 (a) This sounds like a fairly well-designed experiment, so the results should at least apply to this farmer's breed of chicken. **(b)** It would be good to know what proportion of the total student body falls in each of these groups—that is, is anyone overrepresented in this sample? **(c)** How well a TA teaches one topic (power calculations) might not reflect that TA's overall effectiveness.

12.19 (a) With $I = 5$ and $N = 183$, we have df 4 and 178. **(b)** If the reported df were correct (meaning 32 athletes dropped out of the study), 151 athletes actually participated. **(c)** Answers will vary. For example, the individuals could have been outliers in terms of their ability to withstand the water-bath pain. In either case of low or high outliers, their removal would lessen the standard deviation for their sport and move that sports mean (removing a high outlier would lower the mean, and removing a low outlier would raise the mean).

12.21 (a) If we believe the average score for the four off-task behaviors is less than the paper-and-pencil control, a contrast would be $\psi = \mu_7 - 0.25(\mu_1 + \mu_2 + \mu_3 + \mu_4)$, keeping the convention that we believe the contrast will be positive. **(b)** $H_0 : \psi = 0$ and $H_a : \psi > 0$. **(c)** $\hat{\psi} = 0.62333 -$ $0.25(0.62667 + 0.57 + 0.53667 + 0.53667) = 0.05583$. The standard error is

$$SE_{\hat{\psi}} = 0.1205 \sqrt{\frac{1}{21} + (0.25)^2 \left(\frac{1}{21} + \frac{1}{20} + \frac{1}{20} + \frac{1}{21} \right)} = 0.0295.$$ The test statistic is $t = 0.05583/0.0295 =$

1.894 with *P*-value 0.0302. We can conclude that students using the paper-and-pencil "on task" behavior do significantly better on quizzes like these than students who are off-task using digital technologies.

12.23 (a) The standard deviations meet the rule of thumb to be able to pool variances because $0.622 < 2(0.573) = 1.146$. **(b)** The histograms do not look Normal. The control group has a left-skewed distribution, and the organic food group has a distribution where the extreme values are most prevalent. However, there are at least 20 observations per group, so means may be Normally distributed.

Food	Mean	St. Dev.
Comfort	4.887	0.573
Control	5.082	0.622
Organic	5.584	0.594

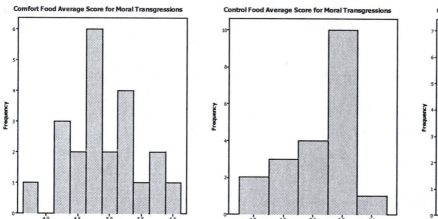

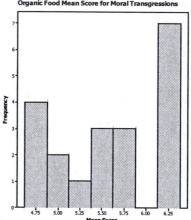

12.25 (a) Student responses will vary, but they should indicate that those with the Comfort prompt were willing to give the fewest number of minutes, on average. Organic appears to differ from both Comfort and Control; Comfort and Control are not significantly different from each other. **(b)** The decrease in variability for the three groups and the curve in the Normal quantile plot might make us question Normality.

12.27 (a) With $I = 3$ and $N = 120$, we have df 2 and 117. **(b)** To use Table E, we compare to df 2 and 100; with $F > 5.02$, we conclude that $P < 0.001$. Software gives $P = 0.0003$. **(c)** Haggling and bargaining behavior is probably linked to the local culture, so we should hesitate to generalize these results beyond similar informal shops in Mexico.

12.29 (a) With roughly equal means, we have $F = 0.00$ and $P = 0.9995$. **(b)** As the means become more different, F increases and P decreases. Here, we have $F = 55.53$ and $P < 0.0001$.

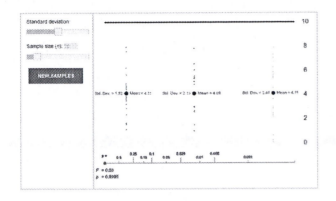 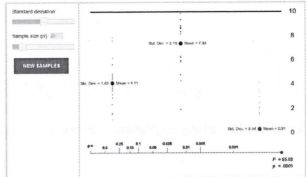

12.31 (a) The data are summarized below.

Group	n	$\overline{x}$	s
Control	35	−1.01	11.50
Group	34	−10.79	11.14
Individual	35	−3.71	9.08

(b) Yes, 2(9.08) = 18.16 > 11.50. **(c)** Histograms follow. Control is closest to a symmetric distribution; individual seems left-skewed. However, with sample sizes at least 34 in each group, moderate departures from Normality are not a problem.

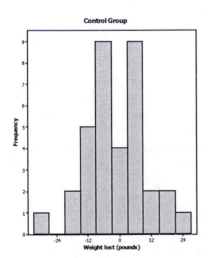 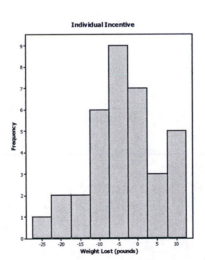 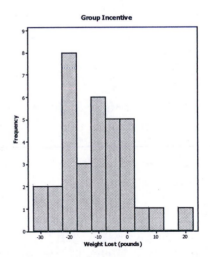

12.33 (a) All weight loss figures were divided by 2.2. The first individual (who actually gained 5.3 pounds) has now gained 2.4091 kg. **(b)** The F statistic and P-value are given with the ANOVA table on the next page. Neither of these has changed, although the sums of squares and mean squares are different. The new values of SS are the previous ones divided by 2.2^2.

Minitab Output: KG versus Group

```
Source    DF      SS      MS      F       P
Group      2    362.1   181.1   7.77   0.001
Error    101   2354.1    23.3
Total    103   2716.2

S = 4.828    R-Sq = 13.33%    R-Sq(adj) = 11.62%
```

12.35 (a) Based on the sample means, fiber is cheapest and cable is most expensive. (Note that the providers are shown in this plot in alphabetical order, but they can be rearranged in any order.) **(b)** Yes; the smallest-to-largest standard deviation ratio is $\dfrac{40.39}{26.09} = 1.55 < 2$. **(c)** The degrees of freedom are $I - 1 = 2$ and $N - I = 44$. From Table E (with df 2 and 40), we have $0.025 < P < 0.050$; software gives $P = 0.0427$. The difference in means is (barely) significant at the 5% level.

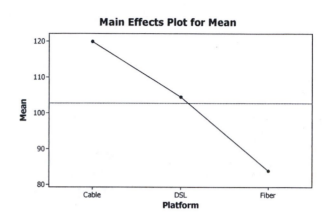

12.37 (a) The variation in sample size is some cause for concern, but there can be no extreme outliers in a 1-to-7 scale, so ANOVA is probably reliable. **(b)** Pooling is reasonable: $1.26/1.03 = 1.22 < 2$. **(c)** With $I = 5$ groups and total sample size $N = 410$, we use an $F(4, 405)$ distribution. We can compare 5.69 to an $F(4, 200)$ distribution in Table E and conclude that $P < 0.001$, or with software determine that $P = 0.0002$. **(d)** Hispanic Americans have the highest emotion scores, Japanese are in the middle, and the other three cultures are the lowest (and very similar).

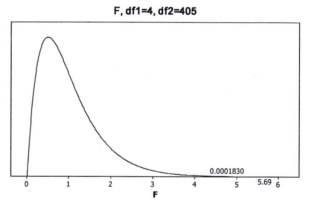

12.39 (a) The plot suggests that both drugs cause an increase in activity level, and Drug B appears to have a greater average effect. **(b)** Yes: The guidelines for pooling standard deviations say that the ratio of largest to smallest should be less than 2; we have $\sqrt{\dfrac{17.20}{7.75}} = 1.49 < 2$. The pooled variance is

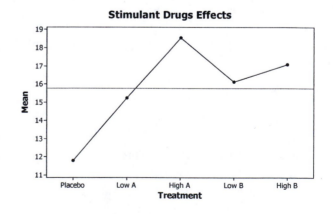

$$s_p^2 = \frac{(5-1)(17.2)+(5-1)(13.1)+(5-1)(10.25)+(5-1)(7.75)+(5-1)(12.5)}{25-5} = 12.16 \text{ and } s_p = 3.487. \textbf{(c)}$$

The degrees of freedom are DFG $= I - 1 = 4$ and DFE $= N - I = 20$. **(d)** Comparing to an $F(4, 20)$ distribution in Table E, we see that $2.25 < F < 2.87$, so $0.05 < P < 0.10$; software gives $P = 0.0642$. We do not have significant evidence of a difference in mean effect.

12.41 Because the descriptions of these contrasts do not specify an expected direction for the comparison, the subtraction could be done either way (in the order shown, or in the opposite order). **(a)** $\psi_1 = \mu_2 - 0.5(\mu_1 + \mu_4)$. **(b)** $\psi_2 = \frac{1}{3}(\mu_1 + \mu_2 + \mu_4) - \mu_3$.

12.43 (a) Pooling is reasonable: The ratio is $0.824/0.657 = 1.25$. For the pooled standard deviation, we compute

$$s_p^2 = \frac{(489-1)(0.804)^2 + (69-1)(0.824)^2 + (212-1)(0.657)^2}{489+69+212-3} = 0.5902 \text{ so } s_p = \sqrt{0.5902} = 0.7683. \textbf{(b)}$$

Comparing $F = 17.66$ to an $F(2, 767)$ distribution, we find $P < 0.001$. Sketches of this distribution will vary; in the graph on the right, we see the 1% critical value is 4.633, so we can see that the observed value lies well above the bulk of this distribution. **(c)** For the contrast $\psi = \mu_2 - 0.5(\mu_1 + \mu_3)$, we test H_0: $\psi = 0$ versus H_a: $\psi > 0$. We find $c = 0.585$ with $SE_c = 0.0977$, so $t = c/SE_c = 5.99$ with df $= 767$, and $P < 0.0001$.

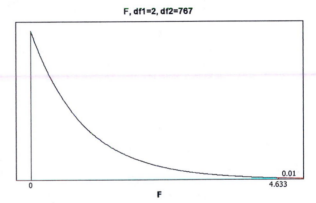

F, df1=2, df2=767

12.45 (a) The means, standard deviations, and standard errors are given (all in grams per cm^2). **(b)** All three distributions appear to be reasonably close to Normal, and the standard deviations are suitable for pooling. **(c)** ANOVA gives $F = 7.72$ (df 2 and 42) and $P = 0.001$, so we conclude that the means are not all the same. **(d)** With df $= 42$, three comparisons, and $\alpha = 0.05$, the Bonferroni critical value is $t^{**} = 2.4937$. The pooled standard deviation is $s_p = 0.01437$, and the standard error of each difference is

$$SE_D = s_p\sqrt{1/15+1/15} = 0.005247,$$ so two means are significantly different if they differ by $t^{**}SE_D = 0.01308$. The high-dose mean is significantly different from the other two. **(e)** Briefly: High doses of kudzu isoflavones increase BMD.

Minitab Output: BMD versus Treatment

Source	DF	SS	MS	F
P				
Treatment	2	0.003186	0.001593	7.72
0.001				
Error	42	0.008668	0.000206	
Total	44	0.011853		

	n	$\bar{x}$	s	$s_{\bar{x}}$
Control	15	0.2189	0.01159	0.002992
Low dose	15	0.2159	0.01151	0.002972
High dose	15	0.2351	0.01877	0.004847

```
S = 0.01437    R-Sq = 26.88%    R-Sq(adj) = 23.39%
```

12.47 (a) Pooling is reasonable, as the largest-to-smallest ratio is about 1.65. **(b)** ANOVA gives $F = 7.98$ (df 2 and 27), for which $P = 0.002$. We reject H_0 and conclude that not all means are equal.

	n	$\bar{x}$	s
Control	10	601.1	27.364
Low jump	10	612.5	19.329
High jump	10	638.7	16.594

One-way ANOVA: density versus group

```
Source  DF    SS     MS     F      P
group    2   7434   3717   7.98  0.002
Error   27  12580    466
Total   29  20013

S = 21.58    R-Sq = 37.14%    R-Sq(adj) = 32.49%
```

12.49 Let μ_1 be the placebo mean, μ_2 and μ_3 be the means for low and high doses of Drug A, and μ_4 and μ_5 be the means for low and high doses of Drug B. Recall from Exercise 12.39 that $s_p = 3.487$. **(a)** The first contrast is $\psi_1 = \mu_1 - 1/2(\mu_2 + \mu_4)$; the second is $\psi_2 = \mu_3 - \mu_2 - (\mu_5 - \mu_4)$. **(b)** The estimated contrasts are $c_1 = 11.80 - 0.5(15.25 + 16.15) = -3.9$ and $c_2 = (18.55 - 15.25) - (17.10 - 16.15) = 2.35$. The respective standard errors are

$$SE_{c_1} = s_p \sqrt{1/4 + 0.25/4 + 0/4 + 0.25/4 + 0/4} = 2.1353 \text{ and}$$

$$SE_{c_2} = s_p \sqrt{0/4 + 1/4 + 1/4 + 1/4 + 1/4} = s_p = 3.487.$$ **(c)** The first contrast is significant ($t_1 = -1.826$, one-sided P-value $= 0.0414$), but the second is not ($t_2 = -0.674$, one-sided P-value 0.2540). We have enough evidence to conclude that low doses increase activity level over a placebo, but we cannot conclude that activity level changes due to increased dosage are different between the two drugs.

12.51 (a) Pooling is risky because $8.66/2.89 = 3 > 2$. **(b)** ANOVA gives $F = 137.94$ (df 5 and 12), for which $P < 0.0005$. We reject H_0 and conclude that not all means are equal.

	n	$\bar{x}$	s
ECM1	3	65.00%	8.6603%
ECM2	3	63.33%	2.8868%
ECM3	3	73.33%	2.8868%
MAT1	3	23.33%	2.8868%
MAT2	3	6.67%	2.8868%
MAT3	3	11.67%	2.8868%

Minitab Output: Gpi versus Material

```
Source    DF      SS       MS      F       P
Material   5   13411.1   2682.2   137.94  0.000
Error     12    233.3     19.4
Total     17   13644.4

S = 4.410    R-Sq = 98.29%    R-Sq(adj) = 97.58%
```

	n	$\bar{x}$
30 min, 1 × /wk	22	−17.4
30 min, 2 × /wk	24	−18.4
60 min, 1 × /wk	24	−24.0
60 min, 2 × /wk	25	−24.0

12.53 For convenience the means and sample sizes are repeated here. In Exercise 12.26, we found $s_p = 18.421$.

Usual care (control)	24	−6.3

(a) We have $\psi_1 = \mu_c - 0.25(\mu_{30\times1} + \mu_{30\times2} + \mu_{60\times1} + \mu_{60\times2})$, $\psi_2 = 0.5(\mu_{30\times1} + \mu_{30\times2})$ $- 0.5(\mu_{60\times1} + \mu_{60\times2})$, $\psi_3 = 0.5(\mu_{60\times1} - \mu_{60\times2}) - 0.5(\mu_{30\times1} - \mu_{30\times2})$.

(b) $c_1 = -6.3 - 0.25(-17.4 - 18.4 - 24 - 24) = 14.65$, $c_2 = 0.5(-17.4 - 18.4) - 0.5(-24.0 - 24.0) = 6.1$, $c_3 = 0.5(-24 - 24) - 0.5(-17.4 - 18.4) = -0.5$. $SE_{c_1} = 4.209$, $SE_{c_2} = SE_{c_3} = 3.784$. **(c)** All tests are two-sided (we are interested only in the comparisons; no direction was given). There were df = 114 in the ANOVA for error. The test results are: $t_1 = 14.65/4.209 = 3.481$, $P < 0.001$ (from Table D, df = 100) and $P = 0.0007$ (from software); $t_2 = 6.1/3.784 = 1.612$, $0.10 < P < 0.20$ (from Table D, df = 100) and $P = 0.1097$ (from software); $t_3 = -0.5/3.784 = -0.132$, $P > 0.5$ (from Table D, df = 100) and $P = 0.8952$ (from software). The average of the massages relieves arthritis pain in the knee better than conventional treatment; there is not enough evidence to conclude that twice a week relieves pain better than once a week; there is no significant difference in the difference of the averages for 30 and 60 minutes. It appears that 60 minutes once or twice a week relieves knee osteoarthritis pain best.

12.55 (a) The plot shows granularity (which varies between groups), but that should not make us question independence; it is due to the fact that the scores are all integers. **(b)** The ratio of the largest to the smallest standard deviations is $1.595/0.931 = 1.713$, which is less than 2. **(c)** Apart from the granularity, the quantile plots (on the following page) are reasonably straight. **(d)** Again, apart from the granularity, the residual quantile plot (below, right) looks pretty good.

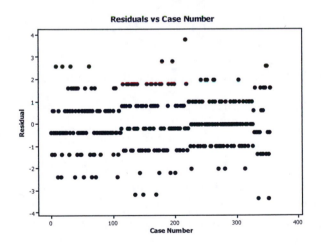

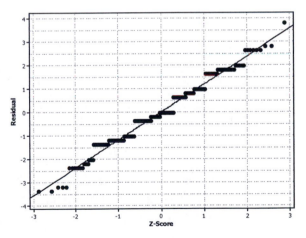

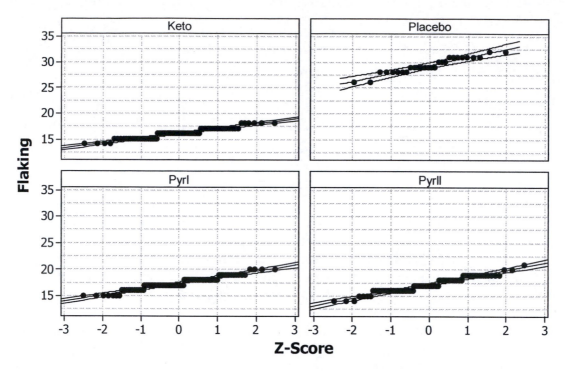

12.57 (a) The three contrasts are

$$\psi_1 = \frac{1}{3}\left(\mu_{Py1} + \mu_{Py2} + \mu_K\right) - \mu_P$$

$$\psi_2 = 0.5\left(\mu_{Py1} + \mu_{Py2}\right) - \mu_K$$

$$\psi_3 = \mu_{Py1} - \mu_{Py2}$$

$c_1 = -12.51$	$c_2 = 1.269$	$c_3 = 0.191$
$SE_{c_1} = 0.2355$	$SE_{c_2} = 0.1413$	$SE_{c_3} = 0.1609$
$t_1 = -53.17$	$t_2 = 8.98$	$t_3 = 1.19$
$P_1 < 0.0005$	$P_2 < 0.0005$	$P_3 = 0.2359$

(b) The pooled standard deviation is $s_p = \sqrt{MSE} = 1.1958$. The estimated contrasts and their standard errors are in the table. For example,

$$SE_{c_1} = s_p\sqrt{\frac{1}{9}/112 + \frac{1}{9}/109 + \frac{1}{9}/106 + 1/28} = 0.2355.$$ **(c)** We test $H_0 : \psi_i = 0$ versus $H_a : \psi_i \neq 0$

for each contrast. The *t*- and *P*-values are given in the table. The Placebo mean is significantly higher than the average of the other three, while the Keto mean is significantly lower than the average of the two Pyr means. The difference between the Pyr means is not significant (meaning the second application of the shampoo is of little benefit)—this agrees with our conclusion from Exercise 12.56.

12.59 Because the measurements in Exercise 12.51 are percents, the instructions to "add 5% to each response" could be interpreted in two ways:
(1) new response = old response + 5
(2) new response = old response × 1.05
The table on the right gives summary statistics for both interpretations (all numbers in percents). For (1), the means increase by 5, but everything else remains

	Version (1)		Version (2)	
	$\overline{x}$	s	$\overline{x}$	s
ECM1	70.0	8.6603	68.25	9.0933
ECM2	68.3	2.8868	66.5	3.0311
ECM3	78.3	2.8868	77.0	3.0311
MAT1	28.3	2.8868	24.5	3.0311
MAT2	11.7	2.8868	7.0	3.0311
MAT3	16.7	2.8868	12.25	3.0311

the same; the ANOVA table (below) is identical to the one in the solution to Exercise 12.51. For (2), both the means and standard deviations are multiplied by 1.05, SS and MS are multiplied by 1.05^2, but F and P remain the same (ANOVA table below).

One-way ANOVA: Gpi+5 versus Material

```
Source     DF      SS      MS       F       P
Material    5  13411.1  2682.2  137.94   0.000
Error      12    233.3    19.4
Total      17  13644.4

S = 4.410   R-Sq = 98.29%   R-Sq(adj) = 97.58%
```

One-way ANOVA: Gpi*1.05 versus Material

```
Source     DF      SS      MS       F       P
Material    5  14785.8  2957.2  137.94   0.000
Error      12    257.3    21.4
Total      17  15043.0

S = 4.630   R-Sq = 98.29%   R-Sq(adj) = 97.58%
```

12.61 A table of means and standard deviations is given. Quantile plots are not shown, but apart from the granularity of the scores and a few possible outliers, there are no marked deviations from Normality. Pooling is reasonable for both PRE1 and PRE2; the ratios are 1.24 and 1.48. For both PRE1 and PRE2, we test $H_0 : \mu_B = \mu_D = \mu_S$ versus H_a : at least one mean is different. Both tests have df 2 and 63. For PRE1, $F = 1.13$ and $P = 0.329$; for PRE2, $F = 0.11$ and $P = 0.895$. There is no reason to believe that the mean pretest scores differ between methods.

Method	n	PRE 1		PRE 2	
		$\overline{x}$	s	$\overline{x}$	s
Basal	22	10.5	2.9721	5.27	2.7634
DRTA	22	9.72	2.6936	5.09	1.9978
Strat	22	9.136	3.3423	4.954	1.8639

Minitab Output: Analysis of Variance on PRE1

```
Source DF      SS     MS     F      p
Group   2   20.58  10.29  1.13  0.329
Error  63  572.45   9.09
Total  65  593.03
```

Analysis of Variance on PRE2

```
Source DF      SS     MS     F      p
Group   2    1.12   0.56  0.11  0.895
Error  63  317.14   5.03
Total  65  318.26
```

12.63 The scatterplot (following, left) suggests that a straight line is *not* the best choice of a model. Regression gives the formula $\widehat{Score} = 4.432 - 0.000102$ Friends. Not surprisingly, the slope is not significantly different from 0 ($t = -0.28$, $P = 0.782$). The regression only explains 0.1% of the variation in score. The residual plot (following, right) is nearly identical to the first scatterplot, and it suggests (as that did) that a quadratic model might be a better choice.

Note: *If one fits a quadratic model, it does better (and has significant coefficients), but it still only explains 8.3% of the variation in attractiveness.*

Minitab Output: Regression of attractiveness score on number of friends
```
The regression equation is
Score = 4.43 - 0.000102 Friends

Predictor        Coef     SE Coef       T       P
Constant       4.4321      0.2060   21.51   0.000
Friends     -0.0001023   0.0003694   -0.28   0.782

S = 1.15028    R-Sq = 0.1%    R-Sq(adj) = 0.0%
```

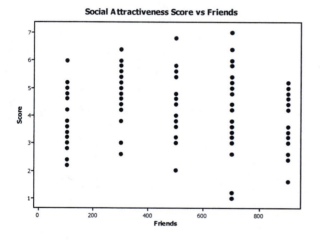 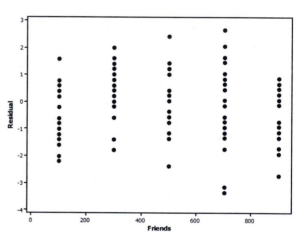

12.65 This exercise asks students to find ANOVA studies on the Internet or in the library.

12.67 (a) Sampling plans will vary but should attempt to address how cultural groups will be determined: Can we obtain such demographic information from the school administration? Do we simply select a large sample then poll each student to determine if he or she belongs to one of these groups? **(b)** Answers will vary with choice of H_a and desired power. For example, with the alternative $\mu_1 = \mu_2 = 4.4$, $\mu_3 = 5$, and standard deviation $\sigma = 1.2$, three samples of size 75 will produce power 0.78. (See Minitab output below.) **(c)** The report should make an attempt to explain the statistical issues involved; specifically, it should convey that sample sizes are sufficient to detect anticipated differences among the groups.

Minitab Output: Power and Sample Size
```
One-way ANOVA
Alpha = 0.05   Assumed standard deviation = 1.2
Factors: 1  Number of levels: 3
```

```
    Maximum   Sample
Difference    Size      Power
      0.6       75    0.782579
```

The sample size is for each level.

12.69 The design can be similar, although the types of music might be different. Bear in mind that spending at a casual restaurant will likely be less than at the restaurants examined in Exercise 12.40; this might also mean that the standard deviations could be smaller. A pilot study might be necessary to get an idea of the size of the standard deviations. Decide how big a difference in mean spending you would want to detect, then do some power computations.

Chapter 13 Solutions

13.1 (a) Two-way ANOVA is used when there are two factors (explanatory variables). (The outcome [response] variable is assumed to have a Normal distribution, meaning that it can take any value, at least in theory.) **(b)** Each level of A should occur with all three levels of B. (Level A has two factors.) **(c)** The RESIDUAL part of the model represents the error. **(d)** DFAB = $(I-1)(J-1)$.

13.3 (a) A *large* value of the AB F statistic indicates that we should reject the hypothesis of no interaction. **(b)** The relationship is backwards: *Mean* squares equal the *sum of* squares divided by degrees of freedom. **(c)** Under H_0, the ANOVA test statistics have an F distribution. **(d)** If the sample sizes are not the same, the sums of squares may not add for "some methods of analysis." (See the "Caution" on page 703; for more detail, see http://afni.nimh.nih.gov/sscc/gangc/SS.html.)

13.5 (a) There are 6 cells with 6 observations each, so $N =$ 36. DFA = 2, DFB = 1, DFAB = 2, DFE = 30. **(b)** See the plot at right. **(c)** $F =$ 2.23 is located at the heavier line in the distribution, P > 0.10 (technology gives $P =$ 0.1251). **(d)** Interaction is not significant; the interaction plot should have roughly parallel lines.

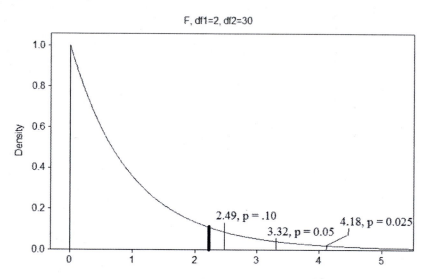

13.7 (a) The factors are gender ($I = 2$) and age ($J = 3$). The response variable is the percent of pretend play. The total number of observations is $N = (2)(3)(11) = 66$. **(b)** The factors are weeks after harvest ($I = 5$) and amount of water ($J = 2$). The response variable is the percent of seeds germinating. The total number of observations is $N = 30$ (3 lots of seeds in each of the 10 treatment combinations). **(c)** The factors are mixture ($I = 6$) and freezing/thawing cycles ($J = 3$). The response variable is the strength of the specimen. The total number of observations is $N =$ 54. **(d)** The factors are training programs ($I = 4$) and the number of days to give the training ($J =$ 2). The response variable is not specified, but presumably is some measure of the training's effectiveness. The total sample size is $N = 80$.

13.9 (a) The same students were tested twice. **(b)** The interactions plot shows a definite interaction; the control group's mean score decreased, while the expressive writing group's mean increased. **(c)** No. $14.3/5.8 = 2.47 > 2$.

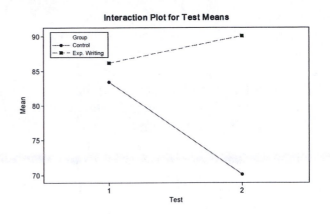

Interaction Plot for Test Means

13.11 (a) The Normality assumption is for the error terms, not the measurements; however, recall from Chapter 12 that ANOVA is robust to reasonable departures from Normality, especially when sample sizes are similar. **(b)** Yes, but barely. $1.62/0.82 = 1.98 < 2$. The ANOVA table is below. Based on this information, Age and Gender are significant factors at the 0.05 level, but the interaction is not significant.

Source	DF	SS	MS	F	P
Age	6	31.97	5.328	4.400	0.0003
Gender	1	44.66	44.66	36.879	0.0000
Age*Gender	6	13.22	2.203	1.819	0.0962
Error	232	280.95	1.211		

13.13 (a) Interaction seems to be present. The means are similar for individuals with a long history with both thank-you conditions; the mean for individuals with a short history is much lower when they were not thanked. **(b)** The marginal mean for short history is 6.245; for long history, the mean is 7.45. People with long histories have a higher repurchase intent. The marginal mean for those who were thanked is 7.085; for those who were not thanked it is 6.61. People who were thanked have a higher mean intention to repurchase than those who were not. This information is useful, but perhaps less so due to the interaction.

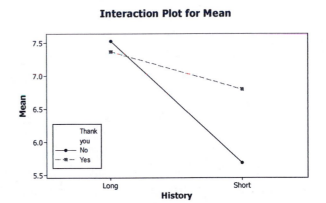

Interaction Plot for Mean

13.15 (a) The plot suggests a possible interaction. (Note that we could have chosen to put proximity type on the horizontal axis instead of dish type; either explanatory variable will do.) **(b)** By subjecting the same individual to all four treatments, rather than four individuals to one treatment each, we reduce the within-groups variability (the residual), which makes it easier to detect between-groups variability (the main effects and interactions).

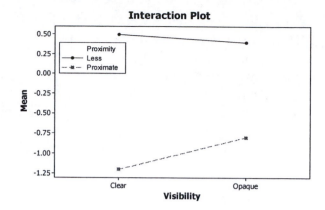

13.17 (a) We'd expect reaction times to slow with older individuals. If bilingualism helps brain functioning, we would not expect that group to slow as much as the monolingual group. The expected interaction is seen in the plot; mean total reaction time for the older bilingual group is much less than for the older monolingual group; the lines are not parallel. **(b)** The interaction is just barely not significant ($F = 3.67$, $P = 0.059$). Both main effects are significant ($P = 0.000$).

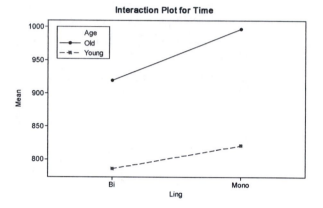

Minitab Output: Time versus Age, Ling

```
Source        DF      SS       MS       F       P
Age            1   479880   479880   195.01   0.000
Ling           1    63394    63394    25.76   0.000
Interaction    1     9031     9031     3.67   0.059
Error         76   187023     2461
Total         79   739329

S = 49.61    R-Sq = 74.70%    R-Sq(adj) = 73.71%
```

13.19 (a) There may be an interaction: For a favorable process, a favorable outcome increases satisfaction quite a bit more than for an unfavorable process (+2.32 versus +0.24). **(b)** With humor, the increase in satisfaction from a favorable outcome is less for a favorable process (+0.49 compared to +1.32). **(c)** There seems to be a three-factor interaction because the interactions in parts (a) and (b) are different.

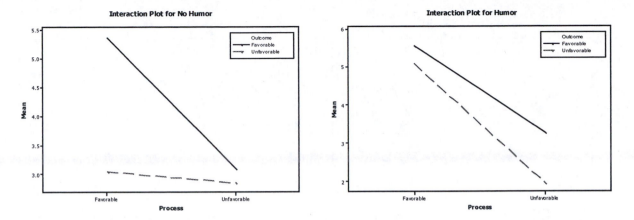

13.21 Marginal means are listed below. In each case, we find the average of the four means for each level of the characteristic. For example, for humor, we have

$$\text{No humor}: \quad \frac{3.04 + 5.36 + 2.84 + 3.08}{4} = 3.58$$

$$\text{Humor}: \quad \frac{5.06 + 5.55 + 1.95 + 3.27}{4} = 3.9575$$

The presence of humor slightly increases mean satisfaction. The process and outcome effects appear to be greater (that is, the change in mean satisfaction is greater).

Humor		Process		Outcome	
No	3.58	Favorable	4.75	Favorable	4.32
Yes	3.96	Unfavorable	2.79	Unfavorable	3.22

13.23 We first find $s_p^2 = \dfrac{(238-1)(1.668)^2 + (125-1)(1.909)^2 + \cdots + (87-1)(1.875)^2}{238 + 125 + \cdots + 87 - 6} = \dfrac{2535.19}{805} = 3.1493$, so

$s_p = \sqrt{3.1493} = 1.7746$. The largest-to-smallest standard deviation ratio is $\dfrac{2.024}{1.601} = 1.26 < 2$, so it

is reasonable to use this pooled estimate.

13.25 Except for female responses to purchase intention, means for the United States and France are less than those in Canada. Females had higher means in almost every case, except for French responses to credibility and purchase intention (suggesting a modest interaction). Gender differences in France are considerably smaller than in either Canada or the United States.

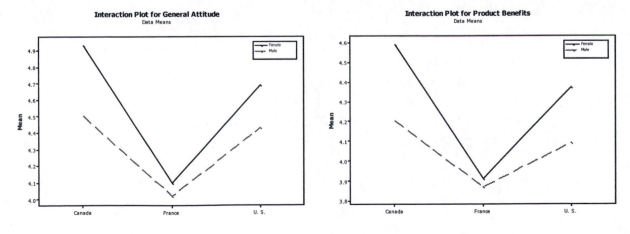

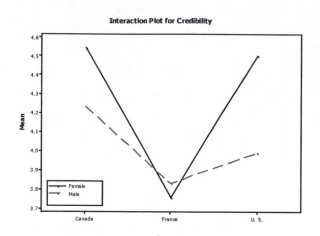

13.27 (a) The marginal means (as well as the individual cell means) are in the table following. The first two means suggest that the intervention group showed more use of safety behaviors than the control group. **(b)** Interaction means that the mean number of behaviors changed differently over time for the two groups. We see this in the plot because the lines connecting the means are not parallel. The intervention group used fewer safety behaviors at the 6-month follow-up than they did at the 3-month follow-up, while the number of behaviors used by the control group continued to increase.

Group	Baseline	3 mo.	6 mo.	Mean
Intervention	10.4	12.5	11.9	11.6
Control	9.6	9.9	10.4	9.967
Mean	10.0	11.2	11.15	10.783

13.29 With $I = 3$, $J = 2$, and 6 observations per cell, we have DFA = 2, DFB = 1, DFAB = 2, and DFE = 30. $3.32 < 3.45 < 4.18$, so $0.025 < P_A < 0.05$ (software gives 0.0448). $2.49 < 2.88$, so $P_B > 0.10$ (software gives 0.1251). $1.14 < 2.49$, so $P_{AB} > 0.10$ (software gives 0.3333). The only significant effect is the main effect for factor A.

13.31 (a) The means are nearly parallel, and show little evidence of an interaction. **(b)** With equal sample sizes, the pooled variance is simply the unweighted average of the variances:
$s_p^2 = 0.25(0.12^2 + 0.14^2 + 0.12^2 + 0.13^2) = 0.016325$.
Therefore, $s_p = \sqrt{0.016325} = 0.1278$.

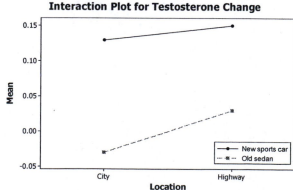

(c) Note that all of these contrasts have been arranged so that, if the researchers' suspicions are correct, the contrast will be positive. To compare new-car testosterone change to old-car change, the appropriate contrast is

$$\psi_1 = 1/2(\mu_{new,city} + \mu_{new,highway}) - 1/2(\mu_{old,city} + \mu_{old,highway})$$

To compare city change to highway change for new cars, we take

$$\psi_2 = \mu_{new,city} - \mu_{new,highway}$$

To compare highway change to city change for old cars, we take

$$\psi_3 = \mu_{old,highway} - \mu_{old,city}$$

(d) By subjecting the same individual to all four treatments, rather than four individuals to one treatment each, we reduce the within-groups variability (the residual) by eliminating the natural differences in testosterone levels among the men, which makes it easier to detect the main effects and interactions of interest.

13.33 (a) Plot on the right. **(b)** There seems to be a fairly large difference between the means based on how much the rats were allowed to eat, but not very much difference based on the chromium level. There may be an interaction: the NM mean is lower than the LM mean, while the NR mean is higher than the LR mean. **(c)** The marginal means are L: 4.86, N: 4.871, M: 4.485, R: 5.246. For low chromium level (L), R minus M is 0.63; for normal chromium (N), R minus M is 0.892. Mean GITH levels are lower

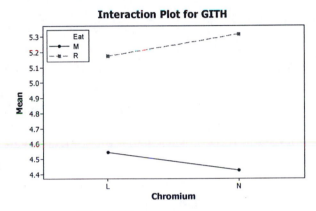

for M than for R; there is not much difference for L versus N. The difference between M and R is greater among rats who had normal chromium levels in their diets (N).

13.35 (a) The pooled variance is $s_p^2 = \dfrac{(32-1)(36.4)^2 + (25-1)(31.2)^2 + (25-1)(41.6)^2 + (27-1)(42.4)^2}{32+25+25+27-4} =$

$\dfrac{152,711.52}{105} = 1454.4$, so $s_p = \sqrt{1454.4} = 38.14$. There were $N = 109$ items in the sample, and four groups, so df $= 105$. **(b)** Pooling is reasonable. The ratio of the largest and smallest standard deviations is $42.4/31.2 = 1.36 < 2$. **(c)** The marginal means are

Sender		Responder	
Individual:	0.5(65.5 + 76.3) = $70.90	Individual:	0.5(65.5 + 54.0) = $59.75
Group:	0.5(54.0 + 43.7) = $48.85	Group:	0.5(76.3 + 43.7) = $60.00

(d) There appears to be an interaction: Individuals send more money to groups, while groups send more money to individuals. **(e)** Use an $F(1, 105)$ distribution. The three P-values are 0.0033 (sender), 0.9748 (responder), and 0.1522 (interaction). Only the main effect of sender is significant.

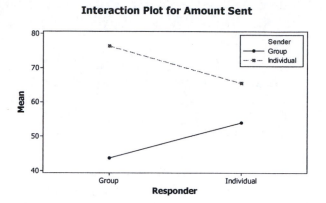

13.37 Yes; the iron-pot means are the highest, and the F statistic for testing the effect of the pot type is very large. (In this case, the interaction does not weaken any evidence that iron-pot foods contain more iron; it only suggests that while iron pots increase iron levels in all foods, the effect is strongest for meats.)

13.39 (a) For all tool/time combinations, $n = 3$. Means and standard deviations are given. Note that five cells had no variability ($s = 0$). **(b)** Plot on the right. Except for tool 1, mean diameter is highest at time 2. Tool 1 had the highest mean diameters, followed by tool 2, tool 4, tool 3, and tool 5. **(c)** Minitab output below; all F statistics are highly significant. **(d)** There is strong evidence of a difference in mean diameter among the tools and among the times. There is also an interaction (tool 1's mean diameters changed differently than the other tools).

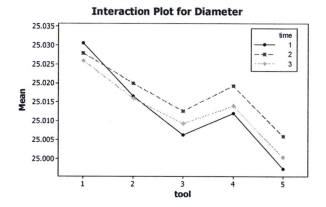

Minitab Output: Two-way ANOVA: diam versus tool, time

Source	DF	SS	MS	F	P
tool	4	0.0035972	0.0008993	412.94	0.000
time	2	0.0001899	0.0000950	43.60	0.000
Interaction	8	0.0001332	0.0000166	7.65	0.000
Error	30	0.0000653	0.0000022		
Total	44	0.0039856			

	Time 1 (8:00AM)		Time 2 (11:00AM)		Time 3 (3:00 PM)	
Tool	$\bar{x}$	s	$\bar{x}$	s	$\bar{x}$	s
1	25.0307	0.001155	25.0280	0	25.0260	0
2	25.0167	0.001155	25.0200	0.002000	25.0160	0
3	25.0063	0.001528	25.0127	0.001155	25.0093	0.001155
4	25.0120	0	25.0193	0.001155	25.0140	0.004000
5	24.9973	0.001155	25.0060	0	25.0003	0.001528

13.41 (a) All three F values have df 1 and 945, and the P-values are < 0.001, < 0.001, and 0.1477. Gender and handedness both have significant effects on mean lifetime, but there is no significant interaction. **(b)** Women live about 6 years longer than men (on the average), while right-handed people average 9 more years of life than left-handed people. "There is no interaction" means that handedness affects both genders in the same way, and vice versa.

13.43 (a and b) The table following lists the means and standard deviations (the latter in parentheses) of the nitrogen contents of the plants. The two plots below suggest that plant 1 and plant 3 have the highest nitrogen content, plant 2 is in the middle, and plant 4 is the lowest. (In the second plot, the points are so crowded together that no attempt was made to differentiate among the different water levels.) There is no consistent effect of water level on nitrogen content. Standard deviations range from 0.0666 to 0.3437, for a ratio of 5.16—larger than we like. **(c)** Minitab output below. Both main effects and the interaction are highly significant.

Minitab Output: Two-way ANOVA: pctnit versus species, water

Source	DF	SS	MS	F	P
species	3	172.392	57.4639	1301.32	0.000
water	6	2.587	0.4311	9.76	0.000
Interaction	18	4.745	0.2636	5.97	0.000
Error	224	9.891	0.0442		
Total	251	189.614			

Species	50mm	150mm	250mm	350mm	450mm	550mm	650mm
1	3.2543	2.7636	2.8429	2.9362	3.0519	3.0963	3.3334
	(0.2287)	(0.0666)	(0.2333)	(0.0709)	(0.0909)	(0.0815)	(0.2482)
2	2.4216	2.0502	2.0524	1.9673	1.9560	1.9839	2.2184
	(0.1654)	(0.1454)	(0.1481)	(0.2203)	(0.1571)	(0.2895)	(0.1238)
3	3.0589	3.1541	3.2003	3.1419	3.3956	3.4961	3.5437
	(0.1525)	(0.3324)	(0.2341)	(0.2965)	(0.2533)	(0.3437)	(0.3116)
4	1.4230	1.3037	1.1253	1.0087	1.2584	1.2712	0.9788
	(0.1738)	(0.2661)	(0.1230)	(0.1310)	(0.2489)	(0.0795)	(0.2090)

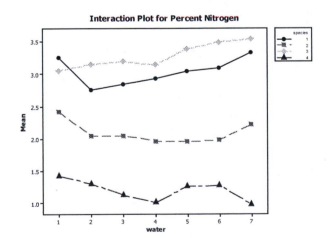

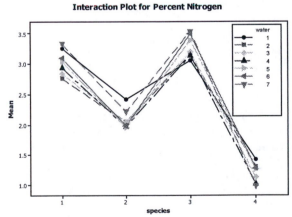

13.45 For each water level, there is highly significant evidence of variation in nitrogen level among plant species (Minitab output below). For each water level, we have df = 32, 6 comparisons, and $\alpha = 0.05$, so the Bonferroni critical value is $t^{**} = 2.8123$. (If we take into account that there are 7 water levels, so that overall we are performing $6 \times 7 = 42$ comparisons, we should take $t^{**} = 3.5579$.) The table on the right gives the pooled

Waterlevel	s_p	SE_D	MSD1	MSD2
1	0.1824	0.0860	0.2418	0.3059
2	0.2274	0.1072	0.3015	0.3814
3	0.1912	0.0902	0.2535	0.3208
4	0.1991	0.0939	0.2640	0.3340
5	0.1994	0.0940	0.2643	0.3344
6	0.2318	0.1093	0.3073	0.3887
7	0.2333	0.1100	0.3093	0.3913

standard deviations s_p, the standard errors of each difference $SE_D = s_p \sqrt{1/9 + 1/9}$, and the "minimum significant difference" MSD = $t^{**} SE_D$ (two means are significantly different if they differ by at least this amount). MSD1 uses $t^{**} = 2.8123$, and MSD2 uses $t^{**} = 3.5579$. As it happens, for either choice of MSD, the only *non*significant differences are between species 1 and 3 for water levels 1, 4, and 7. (These are the three closest pairs of points in the plot from the solution to Exercise 13.39.) Therefore, for every water level, species 4 has the lowest nitrogen level and species 2 is next. For water levels 1, 4, and 7, species 1 and 3 are statistically tied for the highest level; for the other levels, species 3 is the highest, with species 1 coming in second.

Minitab Output: Pct Nitrogen vs Species for Water 1

Source	DF	SS	MS	F	P
species_1	3	18.3711	6.1237	184.05	0.000
Error	32	1.0647	0.0333		
Total	35	19.4358			

Pct Nitrogen vs Species for Water 2

Source	DF	SS	MS	F	P
species_2	3	17.9836	5.9945	115.93	0.000
Error	32	1.6546	0.0517		
Total	35	19.6382			

Pct Nitrogen vs Species for Water 3

Source	DF	SS	MS	F	P
species_3	3	22.9171	7.6390	208.87	0.000
Error	32	1.1704	0.0366		
Total	35	24.0875			

Pct Nitrogen vs Species for Water 4

Source	DF	SS	MS	F	P
species_4	3	25.9780	8.6593	218.37	0.000
Error	32	1.2689	0.0397		
Total	35	27.2469			

Pct Nitrogen vs Species for Water 5

Source	DF	SS	MS	F	P
species_5	3	26.2388	8.7463	220.01	0.000
Error	32	1.2721	0.0398		
Total	35	27.5109			

Pct Nitrogen vs Species for Water 6

Source	DF	SS	MS	F	P
species_6	3	28.0648	9.3549	174.14	0.000
Error	32	1.7191	0.0537		
Total	35	29.7838			

Pct Nitrogen vs Species for Water 7

Source	DF	SS	MS	F	P
species_7	3	37.5829	12.5276	230.17	0.000
Error	32	1.7417	0.0544		
Total	35	39.3246			

13.47 (a and b) The tables below list the means and standard deviations (the latter in parentheses). The means plots show that biomass (both fresh and dry) generally increases with water level for all plants. Generally, species 1 and 2 have higher biomass for each water level, while species 3 and 4 are lower. Standard deviation ratios are quite high for both fresh and dry biomass: $108.01/6.79 = 15.9$ and $35.76/3.12 = 11.5$. **(c)** Minitab output below. For both fresh and dry biomass, main effects and the interaction are significant. (The interaction for fresh biomass has $P = 0.04$; other P-values are smaller.)

Minitab Output: Two-way ANOVA: fresh biomass versus species, water

Source	DF	SS	MS	F	P
species	3	458295	152765	81.45	0.000
water	6	491948	81991	43.71	0.000
Interaction	18	60334	3352	1.79	0.040
Error	84	157551	1876		
Total	111	1168129			

Two-way ANOVA: dried biomass versus species, water

Source	DF	SS	MS	F	P
species	3	50524	16841.3	79.93	0.000
water	6	56624	9437.3	44.79	0.000
Interaction	18	8419	467.7	2.22	0.008
Error	84	17698	210.7		
Total	111	133265			

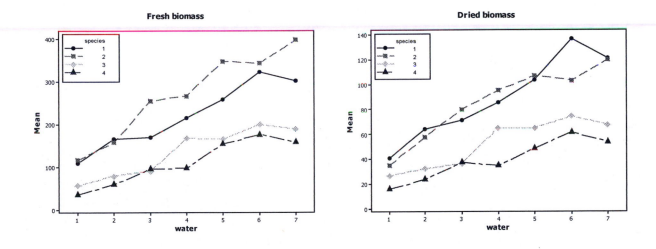

Fresh biomass							
Species	50mm	150mm	250mm	350mm	450mm	550mm	650mm
1	109.095	165.138	168.825	215.133	258.900	321.875	300.880
	(20.949)	(29.084)	(18.866)	(42.687)	(45.292)	(46.727)	(29.896)
2	116.398	156.750	254.875	265.995	347.628	343.263	397.365
	(29.250)	(46.922)	(13.944)	(59.686)	(54.416)	(98.553)	(108.011)
3	55.600	78.858	90.300	166.785	164.425	198.910	188.138
	(13.197)	(29.458)	(28.280)	(41.079)	(18.646)	(33.358)	(18.070)
4	35.128	58.325	94.543	96.740	153.648	175.360	158.048
	(11.626)	(6.789)	(13.932)	(24.477)	(22.028)	(32.873)	(70.105)

			Dry biomass				
Species	50mm	150mm	250mm	350mm	450mm	550mm	650mm
1	40.565	63.863	71.003	85.280	103.850	136.615	120.860
	(5.581)	(7.508)	(6.032)	(10.868)	(15.715)	(16.203)	(17.137)
2	34.495	57.365	79.603	95.098	106.813	103.180	119.625
	(11.612)	(6.149)	(13.094)	(25.198)	(18.347)	(25.606)	(35.764)
3	26.245	31.865	36.238	64.800	64.740	74.285	67.258
	(6.430)	(11.322)	(11.268)	(9.010)	(3.122)	(12.277)	(7.076)
4	15.530	23.290	37.050	34.390	48.538	61.195	53.600
	(4.887)	(3.329)	(5.194)	(11.667)	(5.658)	(12.084)	(25.290)

13.49 For each water level, there is highly significant evidence of variation in biomass level (both fresh and dry) among plant species (Minitab output below). For each water level, we have df = 12, 6 comparisons, and $\alpha = 0.05$, so the Bonferroni critical value is $t^{**} = 3.1527$. (If we take into account that there are 7 water levels, so that overall we are performing $6 \times 7 = 42$ comparisons, we should take $t^{**} = 4.2192$.) The table below gives the pooled standard deviations sp, the standard errors of each difference $SE_D = s_p\sqrt{1/4 + 1/4}$, and the "minimum significant difference" MSD $= t^{**}SE_D$ (two means are significantly different if they differ by at least this amount). MSD1 uses $t^{**} = 3.1527$, and MSD2 uses $t^{**} = 4.2192$. Rather than give a full listing of which differences are significant, we note that plants 3 and 4 are *not* significantly different, nor are 1 and 3 (except for one or two water levels). All other plant combinations are significantly different for at least three water levels. For fresh biomass, plants 2 and 4 are different for *all* levels, and for dry biomass, 1 and 4 differ for all levels.

	Fresh biomass				Dry biomass			
Water level	s_p	SE_D	MSD1	MSD2	s_p	SE_D	MSD1	MSD2
1	20.0236	14.1588	44.6382	50.3764	7.6028	5.3760	16.9487	19.1274
2	31.4699	22.2526	70.1552	79.1735	7.6395	5.4019	17.0305	19.2197
3	19.6482	13.8934	43.8012	49.4318	9.5103	6.7248	21.2010	23.9263
4	43.7929	30.9663	97.6265	110.1762	15.5751	11.0133	34.7213	39.1846
5	38.2275	27.0310	85.2197	96.1746	12.5034	8.8412	27.8734	31.4565
6	59.3497	41.9666	132.3068	149.3147	17.4280	12.3235	38.8518	43.8462
7	66.7111	47.1719	148.7174	167.8348	23.7824	16.8167	53.0176	59.8329

```
Fresh biomass vs Species, Water level 1      Dry biomass vs Species, Water level 1
Source   DF    SS      MS     F      P        Source    DF     SS      MS      F      P
species   3  19107   6369  15.88  0.000       species    3  1411.2  470.4   8.14  0.003
Error    12   4811    401                      Error     12   693.6   57.8
Total    15  23918                             Total     15  2104.8

Fresh biomass vs Species, Water level 2      Dry biomass vs Species, Water level 2
Source   DF    SS      MS     F      P        Source    DF     SS      MS      F      P
species   3  35100  11700  11.81  0.001       species    3  4597.1  1532.4  26.26  0.000
Error    12  11884    990                      Error     12   700.3    58.4
Total    15  46984                             Total     15  5297.4
```

Fresh biomass vs Species, Water level 3						Dry biomass vs Species, Water level 3					
Source	DF	SS	MS	F	P	Source	DF	SS	MS	F	P
species	3	71898	23966	62.08	0.000	species	3	6127.2	2042.4	22.58	0.000
Error	12	4633	386			Error	12	1085.3	90.4		
Total	15	76531				Total	15	7212.6			

Fresh biomass vs Species, Water level 4						Dry biomass vs Species, Water level 4					
Source	DF	SS	MS	F	P	Source	DF	SS	MS	F	P
species	3	62337	20779	10.83	0.001	species	3	8634	2878	11.86	0.001
Error	12	23014	1918			Error	12	2911	243		
Total	15	85351				Total	15	11545			

Fresh biomass vs Species, Water level 5						Dry biomass vs Species, Water level 5					
Source	DF	SS	MS	F	P	Source	DF	SS	MS	F	P
species	3	99184	33061	22.62	0.000	species	3	10026	3342	21.38	0.000
Error	12	17536	1461			Error	12	1876	156		
Total	15	116720				Total	15	11902			

Fresh biomass vs Species, Water level 6						Dry biomass vs Species, Water level 6					
Source	DF	SS	MS	F	P	Source	DF	SS	MS	F	P
species	3	86628	28876	8.20	0.003	species	3	13460	4487	14.77	0.000
Error	12	42269	3522			Error	12	3645	304		
Total	15	128897				Total	15	17105			

Fresh biomass vs Species, Water level 7						Dry biomass vs Species, Water level 7					
Source	DF	SS	MS	F	P	Source	DF	SS	MS	F	P
species	3	144376	48125	10.81	0.001	species	3	14687	4896	8.66	0.002
Error	12	53404	4450			Error	12	6787	566		
Total	15	197780				Total	15	21474			

13.51 (a) With $I = 2$, $J = 3$, and $N = 180$, the degrees of freedom for gender are 1 and 174; for floral characteristic 2 and 174; for interaction 2 and 174. **(b)** Damage to males was higher for all characteristics. For males, damage was highest under characteristic level 3, while for females, the highest damage occurred at level 2. Interaction should be significant because the distance between the means increases from floral type 1 to floral type 3. **(c)** Three of the standard deviations are at least half as large as the means.

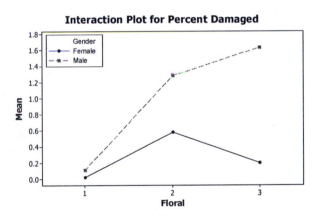

Because the response variable (leaf damage) must be non-negative, this suggests that these distributions are right-skewed; taking logarithms should reduce the skewness.

13.53 The table and plot of the means suggest that females have higher HSE grades than males. For a given gender, there is not too much difference among majors. Normal quantile plots show no great deviations from Normality, apart from the granularity of the grades (most evident among women in EO). In the ANOVA, only the effect of gender is significant. Residual analysis (not shown) reveals some causes for concern; for example, the variance does not appear to be constant.

Gender	CS	EO	Other
Male	$n = 39$	39	39
	$\bar{x} = 7.7949$	7.4872	7.4103
	$s = 1.5075$	2.1505	1.5681
Female	$n = 39$	39	39
	$\bar{x} = 8.8462$	9.2564	8.6154
	$s = 1.1364$	0.7511	1.1611

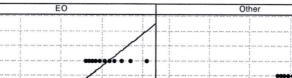

Minitab Output: hse versus Sex, Major

```
Source         DF        SS        MS       F       P
Sex             1   105.338   105.338   50.32   0.000
Major           2     5.880     2.940    1.40   0.248
Interaction     2     5.573     2.786    1.33   0.266
Error         228   477.282     2.093
Total         233   594.073
```

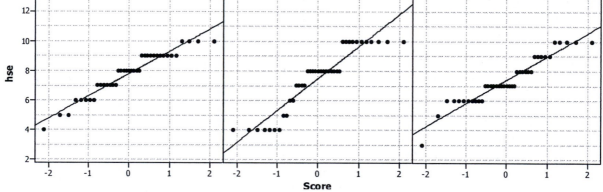

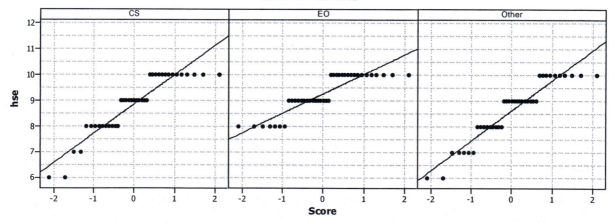

13.55 The following table and plot of the means suggest that students who stay in the sciences have higher mean SATV scores than those who end up in the "Other" group. Female CS and EO students have higher scores than males in those majors, but males have the higher mean in the

"Other" group. Normal quantile plots suggest some right-skewness in the "Women in CS" group and also some non-Normality in the tails of the "Women in EO" group. Other groups look reasonably Normal. In the ANOVA table, only the effect of major is significant.

Gender	CS	EO	Other
Male	$n = 39$	39	39
	$\bar{x} = 526.949$	507.846	487.564
	$s = 100.937$	57.213	108.779
Female	$n = 39$	39	39
	$\bar{x} = 543.385$	538.205	465.026
	$s = 77.654$	102.209	82.184

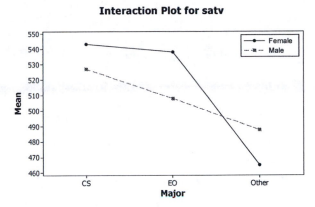

Minitab Output: satv versus Sex, Major

```
Source        DF      SS        MS       F      P
Sex            1     3824    3824.4    0.47   0.492
Major          2   150723   75361.7    9.32   0.000
Interaction    2    29321   14660.7    1.81   0.166
Error        228  1843979    8087.6
Total        233  2027848
```

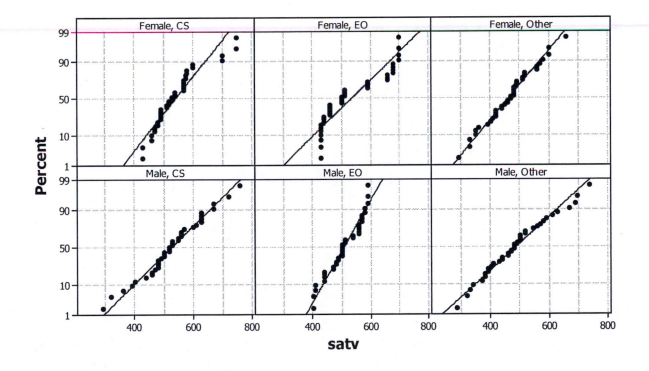

Chapter 14 Solutions

14.1 Odds $= \dfrac{p}{1-p} = \dfrac{1/13}{1-1/13} = \dfrac{1}{12}$, or "1 to 12."

14.3 We have $\hat{p}_{men} = 67/130 = 0.5154$, and $\hat{p}_{women} = 65/140 = 0.4643$. Therefore,

$odds_{men} = \dfrac{0.5154}{1-0.5154} = 1.0636$, and $odds_{women} = \dfrac{0.4643}{1-0.4643} = 0.8667$.

    Note: *The odds can also be computed without first finding $\hat{p}$: for example, 67 men preferred Commercial A, and 63 preferred Commercial B, so $odds_{men} = 67/63$.*

14.5 For the men, $\log(odds) = \ln(1.0635) = 0.0616$; for the women, $\log(odds) = -0.1431$.

    Note: *A student who mistakenly uses the common (base-10) logarithm instead of the natural logarithm will get 0.0267 and −0.0621 as answers.*

14.7 The model is $y = \log(odds) = \beta_0 + \beta_1 x$. If $x = 1$ for men and 0 for women, we need

$\log\left(\dfrac{p_{men}}{1-p_{men}}\right) = \beta_0 + \beta_1$ and $\log\left(\dfrac{p_{women}}{1-p_{women}}\right) = \beta_0$. We estimate $b_0 = \log(odds_{women}) = -0.1431$

and $b_1 = \log(odds_{men}) - b_0 = 0.0616 - (-0.1431) = 0.2047$, so the regression equation is $\log(odds) = -0.1431 + 0.2047x$. If $x = 0$ for men and 1 for women, we estimate $b_0 = \log(odds_{men}) = 0.0616$ and $b_1 = \log(odds_{women}) - b_0 = -0.1431 - 0.0616 = -0.2047$, so the regression equation is $\log(odds) = 0.0616 - 0.2047x$. The estimated odds ratio is either:

$$e^{0.2047} = \dfrac{odds_{men}}{odds_{women}} = 1.2272 \text{ if } x = 1 \text{ for men, or}$$

$$e^{-0.2047} = \dfrac{odds_{women}}{odds_{men}} = 0.8149 \text{ if } x = 1 \text{ for women.}$$

14.9 In Example 14.7, $b_1 = 0.660953$. $e^{0.660953} = 1.936637$, which rounds to 1.94.

14.11 (a) $\hat{p} = 462/1003 = 0.4606$. **(b)** odds $= \dfrac{\hat{p}}{1-\hat{p}} = \dfrac{0.4606}{0.5394} = 0.8539$.

14.13 (a) The model is $y = \log(odds) = \beta_0 + \beta_1 x$. Variable $x = 1$ if the customer is 25 years old or younger and 0 otherwise. **(b)** β_0 is the log(odds) for people over 25 having used their cell phone to consult about a purchase in the last 30 days; β_1 is the difference in log(odds) for those 25 or younger.

14.15 (a) $n = 102{,}738$. Of the recruits, 3869 were rejected and 98,869 were not. **(b)** The equation can be found using the Minitab output: $\log(odds) = -6.76382 + 4.41058x$.

166

14.17 (a) Odds ratio $= e^{4.41058} = 82.3172$. **(b)** The confidence interval is $\left(e^{b_1 - z^* SE_{b_1}}, e^{b_1 + z^* SE_{b_1}}\right) =$
64.7468 to 104.65572. **(c)** Recruits over 40 were about 82.3 times more likely to be rejected for service due to bad teeth than recruits younger than 20. Further, we are 95% confident that recruits 40 years or older were between 64.7 and 104.7 times more likely to be rejected.

14.19 (a) $G = 9413.210$, $P < 0.0005$. We have overwhelming evidence that not all the slopes are 0 (at least one regression coefficient is not 0). **(b)** The confidence interval is $\left(e^{b_1 - z^* SE_{b_1}}, e^{b_1 + z^* SE_{b_1}}\right)$. The results are shown in the table below. **(c)** In each case, the hypotheses are $H_0 : \beta_i = 0$ versus $H_a : \beta_i \neq 0$. The test statistic is $z = b_i / SE_{b_i}$. In all cases, we have $P \approx 0$. In each case we have extremely strong evidence that the slope is not zero; the probability of being rejected for service due to bad teeth is higher for all older age groups than for the under 20 age group.

Age Group	b_1	SE_{b_1}	95% Confidence interval	z
20 to 25	1.97180	0.127593	(5.5941, 9.2247)	15.45
25 to 30	2.85364	0.125048	(13.5793, 22.1399)	22.82
30 to 35	3.55806	0.123713	(27.5384, 44.7252)	28.76
35 to 40	3.96185	0.122837	(41.3094, 66.8606)	32.25
Over 40	4.41058	0.122513	(64.7449, 104.65879)	36.00

14.21 Answers will vary. For example, $68/58{,}952 = 0.0012 = 0.12\%$ of those 20 or younger were rejected. The rejection rate for those 20 to 25 was $647/78{,}639 = 0.0082 = 0.82\%$, $0.82/0.12 = 6.8333$. That analysis would indicate the older recruits were about 6.8333 times more likely to be rejected, while this analysis indicates an odds ratio of 7.1836. The odds ratio is not directly comparable to the ratio of probabilities because it compares odds (the chance of the event happening divided by the chance of the event *not* happening). The first analysis is perhaps more easily understood by most people.

14.23 The odds a student who watches no TV was an exergamer was 0.1249. The odds for a student who reported less than two hours of TV per day is 0.2594. The odds for a student who watches at least two hours of TV per day is 0.4514. The fitted logistic regression equation is $\log(\text{odds}) = -1.34807 + 0.551742(> 2 \text{ hours}) - 0.731368(\text{No TV})$. The odds ratio for no TV to less than 2 hours of TV is $e^{-0.731368} = 0.4813$; the odds ratio for between less than 2 hours of TV and at least 2 hours of TV is $e^{0.551742} = 1.7363$. We can see from the confidence intervals for the slopes that both are significantly different from 0.

Minitab Output: Binary Logistic Regression: X, n versus TV

```
                                              Odds      95% CI
Predictor        Coef     SE Coef        Z       P  Ratio   Lower  Upper
Constant     -1.34807   0.0887320   -15.19   0.000
TV
  > 2 hrs     0.551742   0.143145     3.85   0.000   1.74    1.31   2.30
  No TV      -0.731368   0.442011    -1.65   0.098   0.48    0.20   1.14

Log-Likelihood = -643.018
Test that all slopes are zero: G = 20.193, DF = 2, P-Value = 0.000
```

14.25 (a) The appropriate test would be a chi-square test with df = 4. **(b)** The logistic regression model has no error term. **(c)** H_0 should refer to β_1 (the population slope) rather than b_1 (the estimated slope). **(d)** The interpretation of coefficients is affected by correlations among explanatory variables.

14.27 The regression equation was $\log(\text{odds}) = -3.1658 + 1.3083x$. **(a)** $\log(\text{odds}) = -3.1658 + 1.3083(3.219) = 1.0456$. $e^{1.0456} = 2.845$. **(b)** $\log(\text{odds}) = -3.1658 + 1.3083(3.807) = 1.8149$. $e^{1.8149} = 6.140$. **(c)** $\log(\text{odds}) = -3.1658 + 1.3083(4.174) = 2.2950$. $e^{2.2950} = 9.925$.

14.29 (a) For each column, divide the "yes" entry by the total to find $\hat{p}$. **(b)** For each $\hat{p}$, compute odds $= \hat{p}/(1 - \hat{p})$. **(c)** Finally, take log(odds).

$$\hat{p}_{\text{low}} = \frac{88}{1169} = 0.0753 \quad \text{odds}_{\text{low}} = 0.0753/0.9247 = 0.0814 \quad \log(\text{odds}_{\text{low}}) = -2.5083$$

$$\hat{p}_{\text{high}} = \frac{112}{1246} = 0.0899 \quad \text{odds}_{\text{high}} = 0.0899/0.9101 = 0.0988 \quad \log(\text{odds}_{\text{high}}) = -2.3150$$

14.31 With $b_1 = 3.1088$ and $\text{SE}_{b_1} = 0.3879$, the 99% confidence interval is $b_1 \pm 2.576 \text{SE}_{b_1}$ $= 3.1088 \pm 0.9992$, or 2.1096 to 4.1080.

14.33 (a) $z = \dfrac{3.1088}{0.3879} = 8.01$. **(b)** $z^2 = 64.23$, which agrees with the value of X^2 given by SPSS and SAS. **(c)** The sketches are below. For both the Normal and chi-square distributions, the test statistics are quite extreme, consistent with the reported P-value.

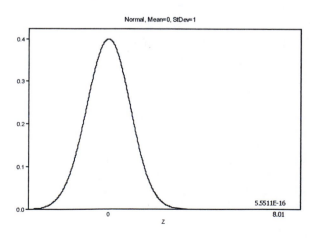
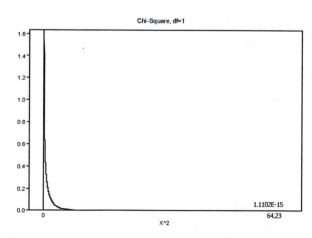

14.35 An odds ratio greater than 1 means a *higher* probability of a *low* tip. Therefore, the odds favor a low tip from senior adults, those dining on Sunday, those who speak English as a second language, and French-speaking Canadians. Diners who drink alcohol and lone males are less likely to leave low tips. For example, for a senior adult, the odds of leaving a low tip were 1.099 (for a probability of 0.5236).

14.37 (a) For the high blood pressure group, $\hat{p}_{hi} = \dfrac{55}{3338} = 0.01648$, giving $\text{odds}_{hi} = \dfrac{\hat{p}_{hi}}{1 - \hat{p}_{hi}} = 0.01675$, or about 1 to 60. (If students give odds in the form "*a* to *b*," their choices of *a* and *b* might be different.) **(b)** For the low blood pressure group, $\hat{p}_{lo} = \dfrac{21}{2676} = 0.00785$, giving $\text{odds}_{lo} = \dfrac{\hat{p}_{lo}}{1 - \hat{p}_{lo}} = 0.00791$, or about 1 to 126 (or 125). **(c)** The odds ratio is $\text{odds}_{hi} / \text{odds}_{lo} = 2.1176$. Odds of death from cardiovascular disease are about 2.1 times greater in the high blood pressure group.

14.39 (a) The estimated odds ratio is $e^{b_1} = 2.1181$ (as we found in Exercise 14.37). Exponentiating the interval for β_1 in Exercise 14.38(a) gives the 95% odds-ratio confidence interval as between 1.28 to 3.51. **(b)** We are 95% confident that the odds of death from cardiovascular disease are about 1.3 to 3.5 times greater in the high blood pressure group than in the low blood pressure group.

14.41 (a) The model is $\log\left(\dfrac{p_i}{1 - p_i}\right) = \beta_0 + \beta_1 x_i$, where $x_i = 1$ if the *i*th person is over 40, and 0 if the person is under 40. **(b)** p_i is the probability that the *i*th person is terminated; this model assumes that the probability of termination depends on age (over/under 40). In this case, that seems to have been the case, but we might expect that other factors were taken into consideration. **(c)** The estimated odds ratio is $e^{b_1} = e^{1.3504} = 3.859$. (Of course, we can also get this from $\dfrac{41/765}{7/504}$.) We can also find, for example, a 95% confidence interval for b_1: $b_1 \pm 1.96\,\text{SE}_{b_1} = 1.3504 \pm 1.96(0.4130) = 0.5409$ to 2.1599. Exponentiating this translates to a 95% confidence interval for the odds: 1.7176 to 8.6701. The odds of being terminated are 1.7 to 8.7 times greater for those over 40. **(d)** Use a multiple logistic regression model, for example,

$$\log \frac{p_i}{1 - p_i} = \beta_0 + \beta_1 x_{1,i} + \beta_2 x_{2,i}.$$

14.43 It is difficult to find the needed probabilities from the numbers as given; this is made easier if we first convert the given information into a two-way table, shown on the right. The proportions meeting the activity guidelines are

	Eats Fruit?		
Active?	Yes	No	Total
Yes	169	494	663
No	68	403	471
Total	237	897	1134

$\hat{p}_{fruit} = 169/237 = 0.7131$ and $\hat{p}_{no} = 494/897 = 0.5507$, so

$\text{odds}_{fruit} = 2.4853$ and $\text{odds}_{no} = 1.2258$. Then, $\log(\text{odds}_{no}) = 0.2036$ and $\log(\text{odds}_{fruit}) = 0.9704$, so

$b_0 = 0.2036$, $b_1 = 0.7068$, and the model is log*(odds)* $= 0.2036 + 0.7068x$. Software reports $SE_{b_1} = 0.1585$ and $z = 4.46$ for testing $H_0 : \beta_1 = 0$. A 95% confidence interval for the odds ratio is 1.49 to 2.77.

Binary Logistic Regression: Active versus Fruit

```
Variable  Value  Count
Active    1        663   (Event)
          0        471
          Total   1134
```

```
Logistic Regression Table
                                                 Odds      95% CI
Predictor       Coef      SE Coef     Z       P   Ratio   Lower   Upper
Constant     0.203599   0.0671244   3.03   0.002
Fruit        0.706792   0.158520    4.46   0.000   2.03    1.49    2.77
```

```
Log-Likelihood = -759.180
Test that all slopes are zero: G = 21.033, DF = 1, P-Value = 0.000
```

14.45 For each group, the probability, odds, and log*(odds)* of being overweight are

$$\hat{p}_{no} = \frac{65{,}080}{238{,}215} = 0.2732 \quad \text{odds}_{no} = \frac{\hat{p}_{no}}{1 - \hat{p}_{no}} = 0.3759 \quad \log(\text{odds}_{no}) = -0.9785$$

$$\hat{p}_{FF} = \frac{83{,}143}{291{,}152} = 0.2856 \quad \text{odds}_{FF} = \frac{\hat{p}_{FF}}{1 - \hat{p}_{FF}} = 0.3997 \quad \log(\text{odds}_{FF}) = -0.9170$$

With $x = 0$ for no fast food and $x = 1$ for fast food, the logistic regression equation is log(odds) $= -0.9785 + 0.0614x$. Software reports $SE_{b_1} = 0.006163$, and for testing $H_0 : \beta_1 = 0$ we have $z = 9.97$, leaving little doubt that the slope is not 0 (the *P*-value is essentially 0). A 95% confidence interval for the odds ratio is 1.0506 to 1.0763; the odds of being overweight for students at schools close to fast-food restaurants are about 1.05 to 1.08 times greater than for students at schools that are not close to fast food.

Binary Logistic Regression: Overweight versus FF nearby

```
Response Information
```

```
Variable     Value    Count
Overweight   1      148223   (Event)
             0      381144
             Total  529367
```

```
Logistic Regression Table
                                                    Odds      95% CI
Predictor        Coef      SE Coef        Z       P  Ratio   Lower   Upper
Constant      -0.978454   0.0045980  -212.80   0.000
FF nearby      0.0614350  0.0061625     9.97   0.000   1.06    1.05    1.08
```

14.47 Minitab output is given below. **(a)** The X^2 statistic for testing this hypothesis is 30.652 (df = 3), which has $P = 0.000$. We conclude that at least one coefficient is not 0. **(b)** The fitted model is log(odds) $= -7.759 + 0.1695$ HSM $+ 0.5113$ HSS $+ 0.1925$ HSE. The standard errors of the three coefficients are 0.1830, 0.2409, and 0.2076, giving respective 95% confidence intervals (using 100 df, $t^* = 1.984$) $0.1695 \pm 1.984(0.183) = -0.1936$ to 0.5326, $0.5113 \pm 1.984(0.2409) = 0.0334$ to 0.9892, and $0.1925 \pm 1.984(0.0.208) = -0.2202$ to 0.6052. **(c)** Only the coefficient of HSS is significantly different from 0.

Note: *The 95% confidence interval given here is for the odds ratio, and not for the coefficient. In the multiple regression case study of Chapter 11, HSM was also the only significant explanatory variable among high school grades, and HSS was not even close to significant.*

Binary Logistic Regression: HIGPA versus HSM, HSS, HSE

Variable	Value	Count	
HIGPA	1	74	(Event)
	0	76	
	Total	150	

Logistic Regression Table

Predictor	Coef	SE Coef	Z	P	Odds Ratio	95% CI Lower	Upper
Constant	-7.75901	1.70465	-4.55	0.000			
HSM	0.169546	0.183021	0.93	0.354	1.18	0.83	1.70
HSS	0.511286	0.240866	2.12	0.034	1.67	1.04	2.67
HSE	0.192453	0.207578	0.93	0.354	1.21	0.81	1.82

Log-Likelihood = -88.633
Test that all slopes are zero: G = 30.652, DF = 3, P-Value = 0.000

Goodness-of-Fit Tests

Method	Chi-Square	DF	P
Pearson	52.7577	53	0.484
Deviance	53.8308	53	0.442
Hosmer-Lemeshow	8.1945	6	0.224

Table of Observed and Expected Frequencies:
(See Hosmer-Lemeshow Test for the Pearson Chi-Square Statistic)

14.49 The coefficients and standard errors for the fitted model are on the following page. Note that the tests requested in parts (a) and (b) are not available with all software packages (such as Minitab). Two packages that will do this are SAS and GLMStat (for Macs).
(a) The X^2 statistic for testing this hypothesis is given by SAS as 18.7174 (df = 3); because $P = 0.0003$, we reject H_0 and conclude that at least one of the high school grades add a significant amount to the model with SAT scores. **(b)** The X^2 statistic for testing this hypothesis is 9.3738 (df = 2); because $P = 0.0092$, we reject H_0; at least one of the two SAT scores adds significantly to the model with high school grades. **(c)** For modeling the odds of HIGPA, high school grades (specifically HSS) are useful; SAT scores (specifically, Math scores) are as well.

Partial SAS output:

Analysis of Maximum Likelihood Estimates

Parameter	higpa	DF	Estimate	Standard Error	Wald Chi-Square	Pr > ChiSq
Intercept	0	1	12.1154	2.4167	25.1315	<.0001
hsm	0	1	-0.00977	0.1970	0.0025	0.9604
hss	0	1	-0.5441	0.2499	4.7412	0.0294
hse	0	1	-0.2777	0.2204	1.5877	0.2077
satm	0	1	-0.00991	0.00343	8.3272	0.0039
satcr	0	1	0.00259	0.00284	0.8324	0.3616

Test that all slopes are zero: G = 40.826, DF = 5, P-Value = 0.000

Linear Hypotheses Testing Results

Label	Wald Chi-Square	DF	Pr > ChiSq
test1_hs	18.7174	3	0.0003
test2_sat	9.3738	2	0.0092

Chapter 15 Solutions

15.1 The data and ranks are shown at right. The ranks for Group A are shaded. They are 1, 2, 3, 4, and 8. The ranks for Group B are 5, 6, 7, 9, and 10.

Group	Rooms	Rank
A	49	1
A	60	2
A	106	3
A	145	4
B	161	5
B	190	6
B	225	7
A	312	8
B	500	9
B	1293	10

15.3 The null hypothesis is H_0: no difference in the distributions of number of rooms. The alternative might be two-sided ("there is a difference") or one-sided if we had a prior suspicion that one group had more rooms than the other. The test statistic is $W = 1 + 2 + 3 + 4 + 8 = 18$.

15.5 Under the null hypothesis, $\mu_W = \dfrac{(5)(11)}{2} = 27.5$ and $\sigma_W = \sqrt{\dfrac{(5)(5)(11)}{12}} = 4.7871$. We found

$W = 18$, so $z = \dfrac{18 - 27.5}{4.7871} = -1.98$, for which the two-sided P-value is $2P(Z \leq -1.98) = 0.0477$.

With the continuity correction, we find $z = \dfrac{18.5 - 27.5}{4.7871} = -1.88$, which gives $P = 2P(Z \leq -1.88) =$

0.0601. The Minitab output on the following page gives a similar P-value to that found with the continuity correction; the difference is due to the rounding of z. In this case, use of the continuity correction changes the decision from "Reject H_0" to "Fail to reject H_0" at the $\alpha = 0.05$ level.

   Note: *If a one-sided alternative was specified in Exercise 15.3, the P-value would be half as large: P = 0.0239, or 0.0301 with the continuity correction.*

   In the Minitab output, the medians are referred to as ETA1 and ETA2 ("eta" is the Greek letter η). Minitab also reports an estimate of -141.0 for the difference $\eta 1 - \eta 2$; note that this is not the same as the difference between the two sample medians (106 − 225). This estimate, called the Hodges-Lehmann estimate, is not discussed in the text and has been removed from the Minitab outputs accompanying other solutions for this chapter. Briefly, this estimate is found by taking every response from the first group and subtracting every response from the second group, yielding (in this case) a total of 25 differences. The median of this set of differences is the Hodges-Lehmann estimate.

Minitab Output: Mann-Whitney Test and CI: A, B

```
     N   Median
A    5   106.0
B    5   225.0

Point estimate for ETA1-ETA2 is -141.0
96.3 Percent CI for ETA1-ETA2 is (-450.9,86.9)
W = 18.0
Test of ETA1 = ETA2 vs ETA1 not = ETA2 is significant at 0.0601
```

15.7 (a) For exergamers, no TV: 6/281= 2.14%, some TV: 160/281 = 56.94%, more than 2 hours TV: 115/281 = 40.93%. For non-exergamers, no TV: 48/919 = 5.22%, some TV: 616/919 = 67.03%, more than 2 hours: 255/919 = 27.75%. **(b)** $\chi^2 = 20.068$, df = 2, $P < 0.0005$. We conclude there is an association between being an exergamer and the amount of TV watching.
Minitab Output: No TV, Some TV, > 2 hrs TV

```
Expected counts are printed below observed counts
Chi-Square contributions are printed below expected counts

        No TV   Some TV   > 2 hrs TV   Total
ExerG      6       160          115     281
        12.65    181.71        86.64
        3.492     2.595         9.282

Not       48       616          255     919
        41.35    594.29        283.36
        1.068     0.793         2.838

Total     54       776          370    1200

Chi-Sq = 20.068, DF = 2, P-Value = 0.000
```

15.9 Rank all the members of the class together. The sorted data and ranks are shown in the table at right. Taking Group 1 to be the women, the sum of their ranks is $3 + 6 + 7 + 9 + 11 = 36$.

Sex	Time	Rank
M	0	1
M	30	2
W	60	3
M	75	4
M	90	5
W	115	6
W	120	7
M	130	8
W	170	9
M	300	10
W	360	11

15.11 As in Exercise 15.9, we take the women to be group 1. $\mu_W = \dfrac{5(11+1)}{2} = 30$ and

$$\sigma_W = \sqrt{\dfrac{(5)(6)(11+1)}{12}} = \sqrt{30} = 5.4772.$$

15.13 (a) From the SAS output, we have $W = 1{,}351{,}159.5$, $z = 8.5063$, $P < 0.0001$. Education and civic participation are related. From the table, we see that as education level increases, individuals are more likely to be engaged. **(b)** For those who are civically engaged, 66.1% have at least some college, and 39.2% are college graduates. The corresponding percents for those who are not civically engaged are 49.7% and 25.9%.

15.15 Back-to-back stemplots (data were rounded to the nearest 1000) are on the right, summary statistics below. The men's distribution is skewed, and the women's distribution has a near-outlier. Men and women are not significantly different ($W = 1421$, $P = 0.6890$). The *t* test assumes Normal distributions; with small samples (like the previous exercise), this might be risky. In this exercise, the samples might be large enough to overcome the apparent non-Normality of the distributions.

Men		Women
3224	0	2
99877555	0	5567888888899
2221100000	1	122244
77765	1	556677777788
44442	2	0111224
76	2	5
10	3	2
6	3	

Note: *Shown below is the Minitab output for a* t *test; the conclusion is the same as the Wilcoxon test (*t = −0.11*, *P = 0.916*).*

	$\overline{x}$	s	Min	Q_1	M	Q_3	Max
Men	14,060	9,065	695	7,464.5	11,118	22,740	36,345
Women	14,253	6,514	2.363	8,345.5	14,602	18,050	32,291

Minitab Output: Wilcoxon test
```
              N   Median
WordsPerDay_1  37   11118
WordsPerDay_2  41   14602
W = 1421.0
Test of ETA1 = ETA2 vs ETA1 not = ETA2 is significant at 0.6890
```
2-sample t test
```
GenderMale1   N    Mean   StDev   SE Mean
1            37   14060    9065      1490
2            41   14253    6514      1017
T-Test of difference = 0 (vs not =): T-Value = -0.11  P-Value = 0.916  DF = 64
```

15.17 (a) Normal quantile plots are not shown. No departures from Normality were seen. **(b)** For testing $H_0 : \mu_1 = \mu_2$ versus $H_a : \mu_1 > \mu_2$, we have $\overline{x}_1 = 0.676$, $s_1 = 0.119$, $\overline{x}_2 = 0.406$, $s_2 = 0.268$. Then, $t = 2.06$, which gives $P = 0.0446$ (df = 5.52). We have fairly strong evidence that high-progress readers have higher mean scores. **(c)** We test: H_0 : Scores for both groups are identically distributed versus H_a : High-progress children systematically score higher for which we find $W = 36$ and $P = 0.0473$. This is the same evidence against H_0, with an essentially identical P-value to that found in part (b).

Minitab Output: High, Low

```
        N   Median
High    5   0.7000
Low     5   0.4000

W = 36.0
Test of ETA1 = ETA2 vs ETA1 > ETA2 is significant at 0.0473
```

15.19 (a) See table. **(b)** For Story 2, $W = 8 + 9 + 4 + 7 + 10 = 38$. Under H_0:

$$\mu_W = \frac{(5)(11)}{2} = 27.5, \quad \sigma_W = \sqrt{\frac{(5)(5)(11)}{12}} =$$

4.7871. **(c)** $z = \dfrac{38 - 27.5}{4.7871} = 2.19$, which gives P $= P(Z > 2.19) = 0.0143$; with the continuity correction, we compute $(37.5 - 27.5)/4.787 = 2.09$, which gives $P = P(Z > 2.09) = 0.0183$. **(d)** See the table.

		Story 1		Story 2	
Child	Progress	Score	Rank	Score	Rank
1	high	0.55	4.5	0.80	8
2	high	0.57	6	0.82	9
3	high	0.72	8.5	0.54	4
4	high	0.70	7	0.79	7
5	high	0.84	10	0.89	10
6	low	0.40	3	0.77	6
7	low	0.72	8.5	0.49	3
8	low	0.00	1	0.66	5
9	low	0.36	2	0.28	1
10	low	0.55	4.5	0.38	2

15.21 We test:

H_0 : Service scores and food scores have the same distribution

versus H_a : Service scores are higher

The differences, and their ranks, are:

Spa	Service score	Food score	Difference	Rank
1	90.6	86.6	4	2
2	87.2	74.4	12.8	7
3	95.0	89.1	5.9	5
4	88.4	81.0	7.4	6
5	91.5	85.7	5.8	4
6	88.2	83.2	5.0	3
7	91.2	93.1	−1.9	1

Here, we note there was only one negative difference (which was the smallest difference in absolute value), so $W+ = 27$.

15.23 Because there are the same number of spas, we still have $\mu_{W^+} = 14$ and $\sigma_{W^+} = 5.9161$.

$z = \dfrac{26.5 - 14}{5.9161} = 2.11$ and $P = 0.0174$.

15.25 $W^+ = 35$.

15.27 $\mu_{W^+} = \dfrac{n(n+1)}{4} = \dfrac{8(9)}{4} = 18$. $\sigma_{W^+} = \sqrt{\dfrac{n(n+1)(2n+1)}{24}} = \sqrt{\dfrac{8(9)(17)}{24}} = 7.1414$.

15.29 (a) $n = 20$. **(b)** $W^+ = 192.0$. **(c)** $P = 0.001$. We reject the null hypothesis of no difference in distributions and conclude one distribution is systematically higher than the other. **(d)** The estimated median is 2.925. With $n = 20$, the median should be the average of the middle two absolute value differences.

Note: *The "estimated median" in the Minitab output (2.925) is not the same as the median of the 20 differences (0.01607). The process of computing this point estimate is not discussed in the text, but we will illustrate it for this simple case: The Wilcoxon estimated median is the median of the set of Walsh averages of the differences. This set consists of every possible pairwise average $(x_i + x_j)/2$ for $i \leq j$; note that this includes $i = j$, in which case the average is x_i. In general, there are $n(n + 1)/2$ such averages, so with $n = 20$ differences, we have 210 Walsh averages.*

15.31 For testing H_0: median $= 0$ versus H_a: median > 0, the Wilcoxon statistic is $W^+ = 119$ (14 of the 15 differences were positive, and the one negative difference was the smallest in absolute value), and $P = 0.001$—very strong evidence that there are more aggressive incidents during Moon days. This agrees with the results of the t and permutation tests.

Minitab Output: Wilcoxon Signed Rank Test: aggdiff

```
Test of median = 0.000000 versus median not = 0.000000

              N for   Wilcoxon            Estimated
          N   Test    Statistic      P     Median
aggdiff  15    15       119.0     0.001     2.570
```

15.33 (a) With this additional subject, six of the seven subjects rated drink A higher, and (as before) the subject who preferred drink B only gave it a 2-point edge. The difference for the seventh subject will be influential for a t test, because its difference (30) is an outlier. **(b)** For testing $H_0: \mu_d = 0$ versus $H_a: \mu_d \neq 0$, we have $\bar{x} = 7.8571$ and $s = 10.3187$, so $t = 2.01$ (df = 6) and $P = 0.091$. **(c)** For testing H_0: Ratings have the same distribution for both drinks versus H_a: One drink is systematically rated higher, we have $W^+ = 26.5$ and $P = 0.043$. **(d)** The new data point is an outlier (the difference of 30 is 2.5 times as large as the next smaller difference of 8), which may make the t procedure inappropriate. This also increases the standard deviation of the differences, which makes t insignificant. The Wilcoxon test is not sensitive to outliers, and the extra data point makes it powerful enough to reject H_0.

Minitab Output: Paired T-Test and CI: DrinkA, DrinkB

```
            N    Mean   StDev  SE Mean
Difference  7    7.86   10.32    3.90
T-Test of mean difference = 0 (vs not = 0): T-Value = 2.01   P-Value = 0.091
```

Wilcoxon Signed Rank Test: C4

```
Test of median = 0.000000 versus median not = 0.000000

            N for   Wilcoxon            Estimated
        N   Test    Statistic      P     Median
C4  7     7         26.5     0.043     5.500
```

15.35 (a) Stemplot at right. The distribution is clearly right-skewed but has no outliers. **(b)** $W^+ = 31$ (only 4 of 12 differences were positive) and $P = 0.556$—there is no evidence that the median is other than 105.

```
 9 | 2
 9 | 578
10 | 024
10 | 55
11 | 1
11 | 9
12 | 2
```

Minitab Output: Wilcoxon Signed Rank Test: radon

```
Test of median = 105.0 versus median not = 105.0

            N for   Wilcoxon
       N    Test    Statistic      P
radon  12    12        31.0    0.556
```

15.37 We want to compare the attractiveness scores for $k = 5$ independent samples (the 102, 302, 502, 702, and 902 friend groups of subjects). Under the null hypothesis for ANOVA, each population is $N(\mu, \sigma)$. An F test is used to compare the group means. The Kruskal-Wallis test only assumes a continuous distribution in each population, and it uses a chi-square distribution for the test statistic.

15.39 Minitab gives the median of each group, and the average rank for each group. Using the "adjusted for ties" values, $H = 17.05$, $P = 0.002$. This indicates that the distributions are not all the same; from the average ranks, the group with the lowest average rank was the 102 friend group; the group with the highest average rank was the 302 friend group.

15.41 (a) For testing

H_0 : The distribution of BMD is the same for all three groups

versus H_a : At least one group is systematically higher or lower

we find $H = 9.12$ with df $= 2$, for which $P = 0.010$. **(b)** In the solution to Exercise 12.45, ANOVA yielded $F = 7.72$ (df 2 and 42) and $P = 0.001$. The ANOVA evidence is slightly stronger, but (at $\alpha = 0.05$) the conclusion is the same.

Minitab Output: Kruskal-Wallis Test: BMD versus treat

```
treat     N   Median   Ave Rank     Z
control   15  0.2190     20.1     -1.05
hidose    15  0.2320     31.2      2.97
lowdose   15  0.2160     17.7     -1.93
Overall   45              23.0

H = 9.10   DF = 2   P = 0.011
H = 9.12   DF = 2   P = 0.010   (adjusted for ties)
```

15.43 (a) Diagram below. **(b)** The stemplots (right) suggest greater density for high-jump rats and a greater spread for the control group. **(c)** $H = 10.68$ with $P = 0.005$. We conclude that bone density differs among the groups. ANOVA tests H_0 : all means are equal, assuming Normal distributions with the same standard deviation. For Kruskal-Wallis, the null hypothesis is that the distributions are the same (but not necessarily Normal). **(d)** There is strong evidence that the three groups have different bone densities; specifically, the high-jump group has the highest average rank (and the highest density), the low-jump group is in the middle,

	Control	Low jump	High jump
55	4		
56	9		
57			
58		8	
59	33	469	
60	03	57	
61	14		
62	1		2266
63		1258	1
64			33
65	3		00
66			
67			4

and the control group is lowest.
Minitab Output: Kruskal-Wallis Test: density versus group

```
group        N  Median  Ave Rank      Z
Control     10   601.5      10.2  -2.33
Highjump    10   637.0      22.6   3.15
Lowjump     10   606.0      13.7  -0.81
Overall     30             15.5


H = 10.66  DF = 2  P = 0.005
H = 10.68  DF = 2  P = 0.005  (adjusted for ties)
```

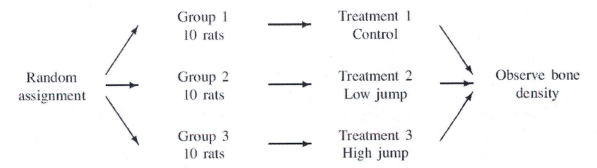

15.45 See also the solution to Exercise 1.88. The means, standard deviations, and medians (all in millimeters) are

Variety	n	$\overline{x}$	s	M
bihai	16	47.5975	1.2129	47.12
red	23	39.7113	1.7988	39.16
yellow	15	36.1800	0.9753	36.11

The appropriate rank test is a Kruskal-Wallis test of H_0 : all three varieties have the same length distribution versus H_a : at least one variety is systematically longer or shorter. We reject H_0 and conclude that at least one species has different lengths ($H = 45.36$, df $= 2$, $P < 0.0005$).

Minitab Output: Kruskal-Wallis Test: length versus variety

```
variety    N  Median  Ave Rank      Z
bihai     16   47.12      46.5   5.76
red       23   39.16      26.7  -0.32
yellow    15   36.11       8.5  -5.51
Overall   54             27.5


H = 45.35  DF = 2  P = 0.000
H = 45.36  DF = 2  P = 0.000  (adjusted for ties)
```

15.47 (a) On the right is a histogram of
service times for Verizon customers.
With only 10 CLEC service calls, it is
hardly necessary to make such a graph
for them; we can simply observe that 7
of those 10 calls took 5 hours, which is
quite different from the distribution for
Verizon customers. The means and
medians tell the same story:

Verizon $\bar{x}_V = 1.7263$ hr $M = 1$ hr

CLEC $\bar{x}_C = 3.80$ hr $M = 5$ hr

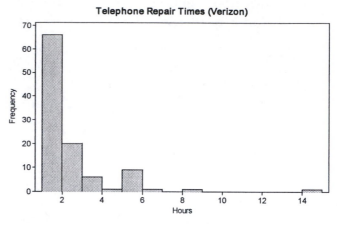

(b) The CLEC distribution has only two
values (1 and 5) with $n = 10$. The t test is not reliable in situations like this. The Wilcoxon rank-
sum test gives $W = 786.5$, which is highly significant ($P = 0.0026$ or 0.0006). We have strong
evidence that response times for CLEC customers are longer. It is also possible to apply the
Kruskal-Wallis test (with two groups). While the P-values are slightly different ($P = 0.0052$, or
0.0012 adjusted for ties), the conclusion is the same: We have strong evidence of a difference in
response times. **(c)** The nonparametric test makes no assumption about the Normality of the
sample means; the mean of the CLEC distribution cannot be Normal with the data we have.

Minitab Output: Mann-Whitney Test and CI: Verizon, CLEC

```
          N   Median
CLEC   10    5.000
ILEC   95    1.000


W = 786.5
Test of ETA1 = ETA2 vs ETA1 > ETA2 is significant at 0.0026
The test is significant at 0.0006 (adjusted for ties)
```

15.49 Use the Wilcoxon rank sum test with a two-sided alternative. For meat, $W = 15$ and
$P = 0.4705$, and for legumes, $W = 10.5$ and $P = 0.0433$ (or 0.0421). There is no evidence
of a difference in iron content for meat, but for legumes the evidence is significant at $\alpha = 0.05$.

Minitab Output: Mann-Whitney Test and CI: Aluminum, Clay for Meat

```
             N   Median
Aluminum   4    2.050
Clay       4    2.375
W = 15.0
Test of ETA1 = ETA2 vs ETA1 not = ETA2 is significant at 0.4705
```

Mann-Whitney Test and CI: Aluminum, Clay for Legumes

```
             N   Median
Aluminum   4    2.3700
W = 10.5
Test of ETA1 = ETA2 vs ETA1 not = ETA2 is significant at 0.0433
The test is significant at 0.0421 (adjusted for ties)
```

15.51 (a) The three pairwise comparisons are *bihai*-red, *bihai*-yellow, and red-yellow. **(b)** The
test statistics and P-values are given in the Minitab output below; all P-values are reported
as 0 to four decimal places. **(c)** All three are easily significant at the overall 0.05 level.

Minitab Output: Wilcoxon rank sum test for *bihai* – red

```
bihai N = 16 Median = 47.120
red   N = 23 Median = 39.160
W = 504.0
Test of ETA1 = ETA2 vs. ETA1 ~= ETA2 is significant at 0.0000
```

Wilcoxon rank sum test for *bihai* – yellow

```
bihai  N = 16 Median = 47.120
yellow N = 15 Median = 36.110
W = 376.0
Test of ETA1 = ETA2 vs. ETA1 ~= ETA2 is significant at 0.0000
```

Wilcoxon rank sum test for red – yellow

```
red    N = 23 Median = 39.160
yellow N = 15 Median = 36.110
W = 614.0
Test of ETA1 = ETA2 vs. ETA1 ~= ETA2 is significant at 0.0000
```

Chapter 16 Solutions

The solutions for Chapter 16 present a special challenge. Because bootstrap and permutation methods require software and random samples, the answers will vary because of (a) random variation due to differences in resampling/rearrangement, and (b) possible systematic and feature differences arising from the specific software used. Because of (a), most of the solutions here give *ranges* of possible answers, rather than a single answer. These ranges should include the results that most students should get from a single bootstrap or permutation run. (Basically, for each such exercises, we reported the minimum and maximum values from 1000 or more bootstraps or permutations.) For (b), the text primarily refers to results from R, but also mentions S+, SAS and SPSS. If you have other statistical software, you can learn about its bootstrapping capabilities (if any) by consulting your documentation, or by doing a Web search for the name of your software and "bootstrap." Note that a free student version of S-PLUS is available from http://estore.onthehub.com/ by clicking the "Data Analysis" tab, then looking for Tibco under Categories, so your students may use it for this chapter, even if they normally work with other software. (Faculty can download a 30-day evaluation copy.) Many of these solutions were originally written by Tim Hesterberg (using S-PLUS) for earlier editions of *IPS,* and have been edited and updated by Pat Humphrey using R (the free, open-source version of S-PLUS, with its Book package) and Minitab. One difference in R's bootstrapping library versus that of S-PLUS is that (at the time of this writing) it does not compute "tilting" confidence intervals, so those results are not given. If your software finds tilting intervals, they will (for most of these exercises) be similar to those found by other methods (percentile, BCa, etc.).

Bootstrapping is not a "native" feature in Minitab, but it can be done easily with an Exec (.mtb) file for straightforward resampling. To bootstrap the distribution of a mean, for example, use an editor such as Notepad to create a file with an .mtb extension. Below is one example (used in the solution to Exercise 16.9).

```
sample 20 c1 c4;
replace.
let c5(k1)=mean(c4)
let k1=k1+1
```

In Minitab, click Editor>Enable commands and initialize the counter k1:

MTB> let k1=1

Run the exec: Click File>Other files>Run an Exec. Enter the number of times to execute the file (here, 2000) and click Select File.

Locate the file you just created (or edited) and click Open to execute it. For the example above, the bootstrap distribution of resample means will be placed in column c5.

16.1 (a) Student answers in this problem will vary substantially due to using different random numbers. (If they do not, you should be suspicious.) **(b)** While students could get a sample mean as low as 3, or as high as 47, 95% of all sample means should be between about 6.5 and 31.5. **(c)** Shown is a stemplot for a set of 200 resamples. Even for such a large number of resamples, the distribution is somewhat irregular; student stemplots (for 20 resamples) will be even more irregular. **(d)** The theoretical bootstrap standard error is about 6.555, but with only 20 resamples, there will be a fair amount of variation (although almost certainly in the range 4.0 to 9.2).

```
0 | 45555
0 | 66666777777
0 | 88999
1 | 000000000001111111111111
1 | 2222222222222233333
1 | 44444444444555555555555
1 | 66666666777777777777777777
1 | 8888888888899999
2 | 0000001111111111
2 | 222223333
2 | 4444444444455555555555
2 | 66666667
2 | 888899999
3 | 0111
3 | 222
3 |
3 | 7
```

 Note: *The range of numbers (6.5 to 31.5) given in part (b) is based on 10,000 resamples. For part (d), the range of standard errors is based on the middle 99% of the SEs from 50,000 separate resamples of size 20. The theoretical value is based on considering the six repair times as a* population *to compute the standard deviation (dividing by* $\sqrt{6}$ *rather than* $\sqrt{5}$*), yielding* $\sigma = 16.0653$*, so the theoretical standard error is* $\sigma/\sqrt{6} = 6.555$*. The computation in the text (page 16-6) does not mention this detail, although it is discussed briefly in Note 3 on page 16-58. Because bootstrap methods are generally not used with small samples, and the difference is negligible for large samples, it usually does not matter.*

16.3 (a) Answers will vary, but a Normal quantile plot indicates that a Normal distribution cannot be rejected. **(b)** Answers will vary. **(c)** The means could range from 24 to 140, however, based on 2000 resamples, the means should range between 33 and 122.

16.5 The mean of the bootstrap distribution is about 75 (indicated by the dashed line below the histogram); it is roughly Normal, but the peak in the center is too high to be truly Normal. The Normal quantile plot is straight, except for some straggling at the low end (and a bit at the high end). Based on these plots, the bootstrap distribution is approximately Normal.

16.7 (a) The standard deviation of the bootstrap distribution will be approximately $s/\sqrt{n}$. **(b)** Bootstrap samples are done *with* replacement from the original sample. **(c)** You should use a sample size *equal* to the original sample size. **(d)** The bootstrap distribution is created by sampling with replacement from the original sample, not the population.

16.9 We bootstrap only the ticket prices (ignoring the number of seats purchased). This distribution is still slightly right-skewed, but close to Normal.

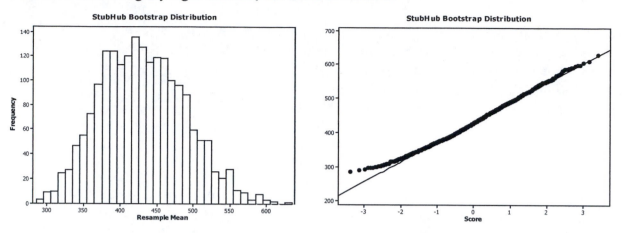

16.11 Starting from a right-skewed distribution, this bootstrap distribution appears approximately

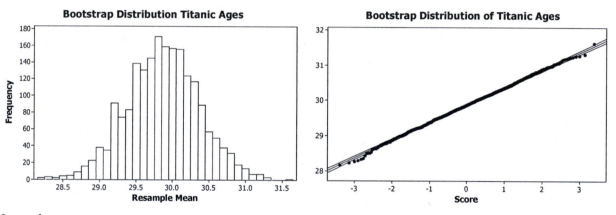

Normal.

16.13 Bootstrap standard errors will vary. **(a)** $s = 14.8$; $s/\sqrt{n} = 14.8/\sqrt{60} = 1.911$. From this author's bootstrap distribution, the standard error is about 1.866. **(b)** $s = 265.150$, so SE = $265.15/\sqrt{20} = 59.289$. This author's bootstrap distribution had a standard deviation of 57.839. **(c)** $s = 3.55887$, so SE = $3.55887/\sqrt{8} = 1.2583$. The author's bootstrap distribution had $s = 1.195$.

16.15 (a) The histogram is shown at right. This distribution looks more strongly skewed than the one in the previous exercise (it would look even more skewed if some of the calls above 600 seconds had been in the sample). We notice one possible outlier about 240 seconds. **(b)** The bootstrap standard error for the sample of 80 calls is about 36.8. For the sample of 10 calls, the standard error is about 29. This is smaller (not larger),

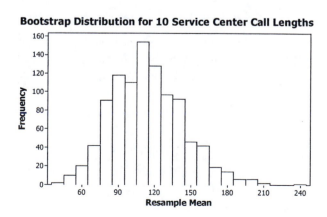

possibly due to the lack of the extremely long calls in the small sample.

16.17 (a) The bootstrap distribution is skewed (refer to the solution to Exercise 16.12); a *t* interval may not be appropriate. **(b)** The bootstrap *t* interval is $\bar{x} \pm t^* SE_{boot}$, where $\bar{x} = 354.1$ sec, $t^* = 2.010$ for df = 49, and SE_{boot} is typically between 39.5 and 46.5. This gives the range of intervals shown on the right. **(c)** The interval reported in Example 7.11 was 266.6 to 441.6 seconds.

Typical ranges	
SE_{boot}	39.5 to 46.5
t lower	260.7 to 274.7
t upper	433.5 to 447.7

16.19 The summary statistics given in Example 16.6 include standard deviations $s_1 = 0.859$ for males and $s_2 = 0.748$ for females, so

$$SE_D = \sqrt{\frac{0.859^2}{91} + \frac{0.748^2}{59}} = 0.1326.$$ The standard error reported by the S-PLUS bootstrap routine in Example 16.6 was 0.1327419 (very close).

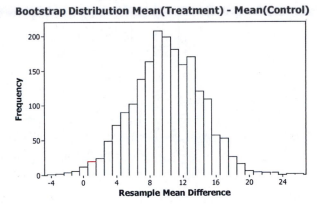

Bootstrap Distribution Mean(Treatment) − Mean(Control)

16.21 With $n = 51$, df = 50, so $t^* = 2.009$. The interval becomes $112.3 \pm (2.009)(9.4) = 112.3 \pm 18.885 = 93.415$ to 131.185.

16.23 (a) $\bar{x} = 29.881$, df = 1045, $t^* = 1.962$, $SE_{boot} = 0.445$, so the interval is 29.008 to 30.754 years. Typical ranges are given at right. **(b)** The standard deviation of the original sample is $s = 14.414$, so the interval is 29.007 to 30.755. The intervals are almost identical, only differing in the third decimal place. **(c)** Answers will vary.

Typical ranges	
SE_{boot}	0.445 to 0.452
t lower	29.006 to 28.994
t upper	30.756 to 30.768

16.25 See also the solution to Exercise 16.14. **(a)** The bootstrap bias is typically between −4 and 4, which is small relative to $\bar{x} = 196.575$ min. **(b)** Ranges for the bootstrap interval are given on the right. **(c)** $SE_{\bar{x}} = 342.022 / \sqrt{80} = 38.2392$, while SE_{boot} ranges from about 35 to 41. The usual *t* interval is 120.46 to 272.69 min.

Typical ranges	
Bias	−4 to 4
SE_{boot}	35 to 41
t lower	114 to 127
t upper	266 to 279

16.27 The bootstrap distribution of the standard deviation shown following looks quite Normal. This particular resample had $SE_{boot} = 0.0489$ and mean 0.814. The original sample of GPAs had $s = 0.817$, so there is little bias. With $n = 150$, we can use df = 100, $t^* = 1.984$ and find a 95% confidence interval for the population standard deviation of $0.817 \pm (1.984)(0.0489) = 0.7120$ to 0.9140.

Typical ranges	
SE_{boot}	0.047 to 0.05
t lower	0.718 to 0.724
t upper	0.910 to 0.916

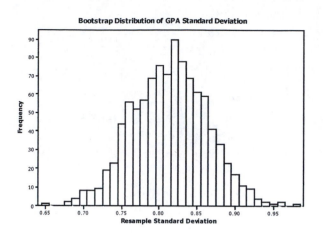

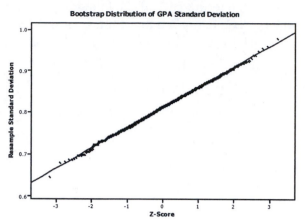

16.29 (a) The data appear to be roughly Normal, though with the typical random gaps and bunches that usually occur with relatively small samples. It appears from both the histogram and quantile plot that the mean is slightly larger than zero, but the difference is not large enough to rule out a $N(0, 1)$ distribution. **(b)** The bootstrap distribution is extremely close to Normal with no appreciable bias.

Typical ranges	
Bias	−0.016 to 0.02
SE_{boot}	0.12 to 0.14
t lower	−0.16 to −0.11
t upper	0.36 to 0.41

(c) $SE_{\bar{x}} = 0.1308$, and the usual t interval is −0.1357 to 0.3854. Typical results for SE_{boot} and the bootstrap interval are above on the right.

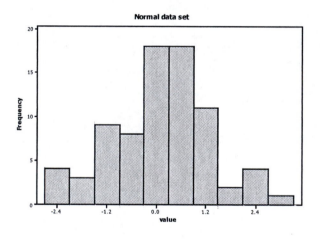

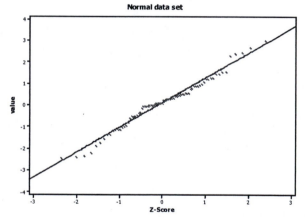

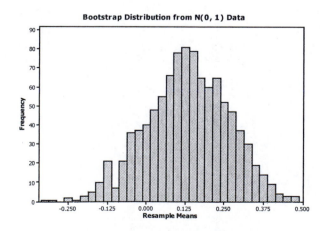

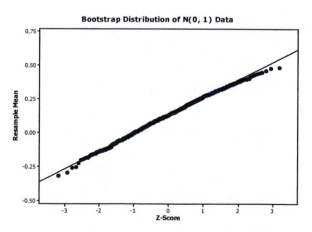

16.31 (a) The sample standard deviation is $s = 4.4149$ mpg. **(b)** The typical range for SE_{boot} is in the table on the right. **(c)** SE_{boot} is quite large relative to s, suggesting that s is not a very accurate estimate. **(d)** There is substantial negative bias and some skewness, so a t interval is probably not appropriate.

Typical ranges	
Bias	-0.22 to -0.09
SE_{boot}	0.55 to 0.65
t lower	3.07 to 3.26
t upper	5.57 to 5.76

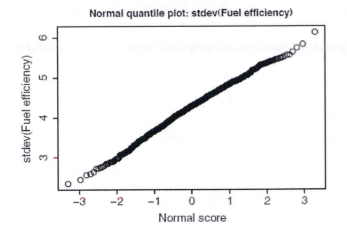

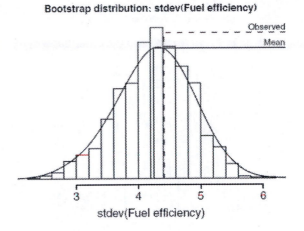

16.33 (a) The distribution of $\bar{x}$ is $N(26, 27/\sqrt{n})$. **(b)** SE_{boot} ranged from about 6.49 to about 7.08 (these values will depend on your original random sample—the author's original sample had a standard deviation of 22.5, which is quite a bit smaller than the population standard deviation). **(c)** For $n = 40$, the original random sample had $s = 29.21$. SE_{boot} ranged from about 4.2 to 4.7. For $n = 160$, SE_{boot} ranged from about 2.02 to about 2.25 (the original sample had $s = 26.73$). **(d)** We note that none of these histograms looks exactly Normal and the decrease in the standard error with increasing n.

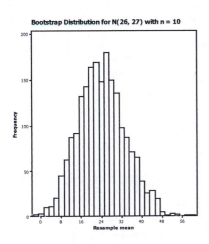

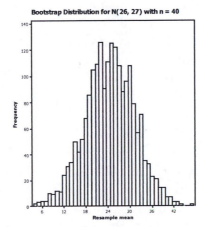

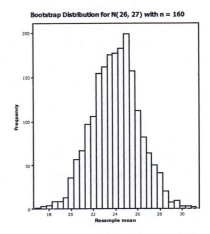

16.35 Student answers should vary depending on their samples. They should notice that the bootstrap distributions are approximately Normal for larger sample sizes. For small samples, the sample could be skewed one way or the other in Exercise 16.33, and most should be right-skewed for Exercise 16.34. Some of that skewness should come through into the bootstrap distribution.

16.37 (a) Both graphs indicate the bootstrap distribution is right-skewed. **(b)** The 95% t confidence interval is $23.26 \pm (2.571)SE_{boot}$. The graphed distribution had $SE_{boot} = 6.71$, so the interval would be 6.0 to 40.5 days. **(c)** The 95% bootstrap percentile confidence interval from the distribution shown is about 6.3 to about 32.2 days, but typical values are from about 6 to 33 days.

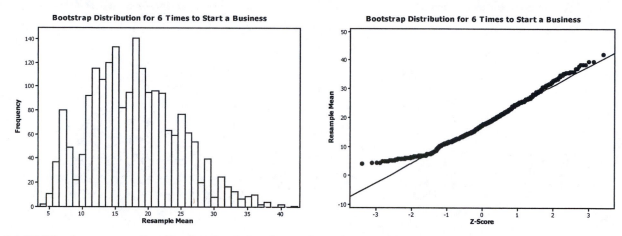

16.39 The bootstrap distribution is right-skewed; an interval based on t would not be appropriate. The original sample had the statistic of interest $\hat{\theta} = 1.21$. The bootstrap distribution had a sample mean a bit higher because the bias is 0.045. The standard deviation of the bootstrap distribution (SE_{boot}) is 0.2336. The BCa confidence interval is (0.766, 1.671) which is located about 0.11 higher than the regular bootstrap interval, which was (0.653, 1.554) and lower than the percentile interval (0.860, 1.762).

16.41 (a) The bootstrap percentile and t intervals are very similar, suggesting that the t intervals are acceptable. **(b)** Every interval (percentile and t) includes 0.

   Note: *In the solution to Exercise 16.29, the percentile intervals were always 70% to 80% as wide as the* t *intervals (because of the heavy tails of that bootstrap distribution). In this case, the width of the percentile interval is 93% to 106% of the width of the* t *interval.*

Typical ranges	
t lower	−0.16 to −0.11
t upper	0.36 to 0.41
Percentile lower	−0.17 to −0.09
Percentile upper	0.35 to 0.42

16.43 The 95% t interval given in Example 16.5 was 2.794 to 3.106. The 95% bootstrap percentile interval is 2.793 to 3.095, as given in Example 16.8. The differences are relatively small relative to the width of the intervals, so they do not indicate appreciable skewness.

16.45 One set of 1000 repetitions gave the BCa interval as (0.4503, 0.8049). We see the bootstrap distribution is left-skewed, and that there was one possible high outlier as well. The lower end of the BCa interval typically varies between 0.42 and 0.46 while the upper end varies between 0.795 and 0.805. These intervals are lower than those found in Example 16.10.

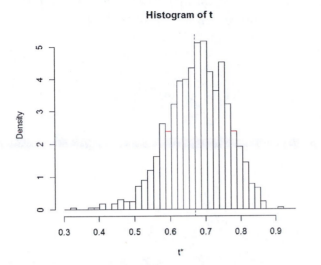

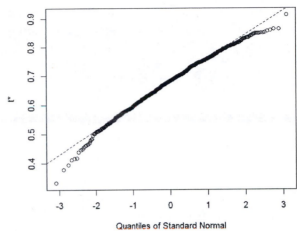

16.47 The percentile interval is shifted to the right relative to the bootstrap *t* interval. The more accurate intervals are shifted even further to the right.

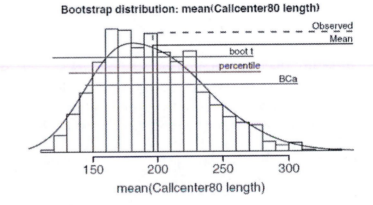

Typical ranges	
t lower	114 to 127
t upper	266 to 279
Percentile lower	127 to 140
Percentile upper	267 to 298
BCa lower	137 to 152
BCa upper	292 to 371

16.49 The interval students will report will vary. Typical ranges are given at right. We note that the locations of the intervals are quite different (especially the low ends); this mirrors what was seen in Exercise 16.47. These intervals are much lower (and narrower) than those in Exercise 16.47.

Typical ranges	
SE_{boot}	27.9 to 29.5
t lower	48 to 52
t upper	178 to 181
Percentile lower	60 to 67
Percentile upper	169 to 177
BCa lower	66 to 72
BCa upper	176 to 195

16.51 Typical ranges for the endpoints of the BCa interval (using males–females) are given at right. This interval is comparable to the −0.416 to 0.118 found in Example 16.6.

Typical ranges	
BCa lower	−0.427 to −0.372
BCa upper	0.087 to 0.133

16.53 (a) The bootstrap distribution is left-skewed (with outliers); simple bootstrap inference is inappropriate. **(b)** The percentile interval is consistently about 0.978 to 0.994, while the BCa interval is consistently 0.970 to 0.993. These agree fairly well, but the BCa interval is shifted right and a bit shorter. We do have significant evidence that the correlation is not 0.

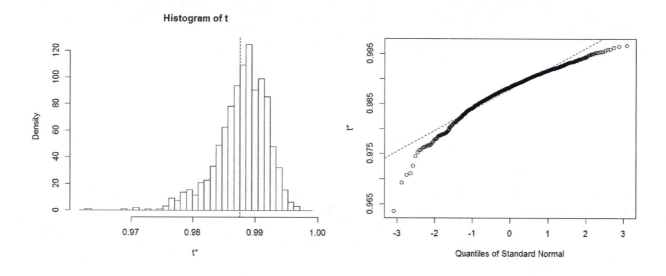

16.55 (a) The regression line is $\widehat{\text{Rating}} = 26.724 + 1.207$ PricePerLoad. **(b)** The ends of the bootstrap distribution do not look very Normal, a t interval may not be appropriate. **(c)** The typical standard error of the slope is 0.2846. With $t_{22} = 2.074$, the typical confidence interval would be $1.207 \pm (2.074)(0.2846)$, or 0.6167 to 1.7973. All these intervals seem to be located higher.

Typical ranges	
SE_{boot}	0.228 to 0.242
t lower	0.705 to 0.734
t upper	1.68 to 1.71
Percentile lower	0.83 to 0.86
Percentile upper	1.74 to 1.81
BCa lower	0.71 to 0.80
BCa upper	1.62 to 1.69

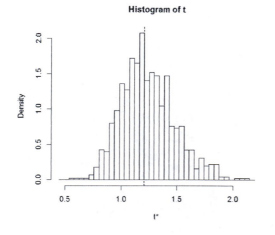
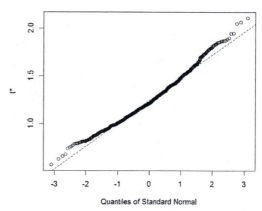

16.57 The regression equation is

$\widehat{Debt2010} = 0.047 + 1.059 Debt2009.$ **(a)** The residuals plots are shown below. The scatterplot indicates a possible increase in variability with increasing 2009 debt, as well as possible outliers (especially the point at upper right, which was Greece). The Normal plot also indicates potential outliers on both ends of the distribution. **(b)** We see indications of high outliers in both the histogram and the Normal quantile plot of the bootstrap distribution, shown on the next page. **(c)** The standard confidence interval for the slope is $1.059 \pm (2.040)(0.03027) = 0.997$ to 1.121. This interval is close to all the bootstrap intervals, but a bit narrower.

Typical ranges	
Bias	−0.003 to −0.007
SE_{boot}	0.043 to 0.047
t interval lower	0.96 to 0.97
t interval upper	1.15 to 1.16
Percentile lower	0.96 to 0.97
Percentile upper	1.13 to 1.14
BCa lower	0.98 to 0.99
BCa upper	1.14 to 1.16

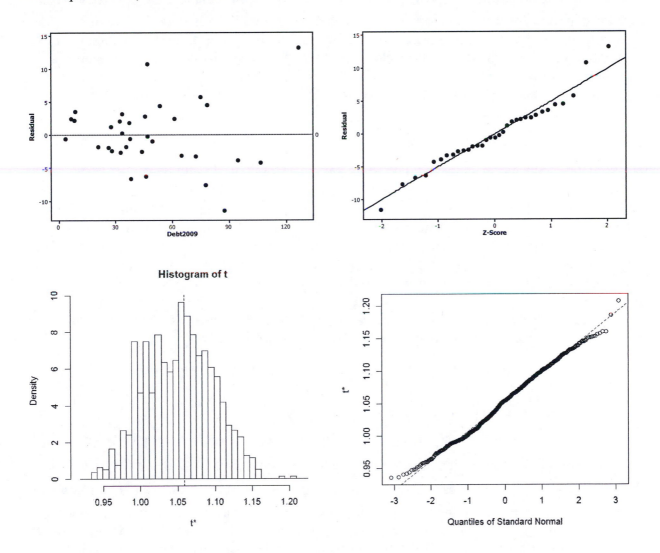

16.59 No, because we believe that one population has a smaller variability. In order to pool the data, the permutation test requires that both populations be the same when H_0 is true.

16.61 Enter the data with the score given to the phone and an indicator for each design. We have hypotheses $H_0 : \mu_1 = \mu_2$ (no difference in preference) and $H_a : \mu_1 \neq \mu_2$ (there is a preference for one of the phones, but we don't know which one). Resample the design indicators (without replacement) to scramble them. Compute the mean score for each scrambled design group. Repeat the process many times. The *P*-value of the test will be the proportion of resamples where the resampled difference in group means is larger than the observed difference (in absolute value).

16.63 If there is no relationship, we have $H_0 : \rho = 0$. We test this against $H_a : \rho \neq 0$ (there is a relationship of unspecified direction). Because there is no relationship under H_0, we can resample one of the variables, say screen satisfaction (without replacement), and compute the correlation between that and the original scores for keyboard satisfaction. Repeat the process many times, keeping track of the proportion of resamples where the correlation is greater in absolute value than that found in the original data. That proportion is the *P*-value for the test.

16.65 (a) The observed difference in means is $\dfrac{57 + 53}{2} - \dfrac{19 + 37 + 41 + 42}{4} = 55 - 34.75 = 20.25$.

(b) Student results will vary, but should be one of the 15 (equally likely) possible values: $-20.25, -17.25, -16.5, -8.25, -5.25, -3.75, -3, 0, 5.25, 8.25, 9, 11.25, 12, 20.25$.
(c) The histogram shape will vary considerably. **(d)** Out of 20 resamples, the number which yield a difference of 20.25 (or more) has a binomial distribution with $n = 20$ and $p = 1/15$, so most students should get between 0 and 4 resamples that give a value of 20.25, for a *P*-value between 0 and 0.2. **(e)** As was noted in part (b), only one resample possibility can give a difference of means greater than or equal to the observed value, so the exact *P*-value is $1/15 = 0.0667$.

    Note: *To determine the 15 possible values, note that the six numbers sum to 249. If the first two numbers add up to* T, *then the other four will add up to 249 − T, and the difference in means will be* $\dfrac{1}{2}T - \dfrac{1}{4}(249 - T) = \dfrac{3}{4}T - 62.25.$ *The values of* T *range from* $19 + 37 = 56$ *to* $57 + 53 = 110$.

16.67 (a) $H_0 : \mu_{EE} = \mu_{LE}$ and $H_a : \mu_{EE} \neq \mu_{LE}$ (the question of interest in Example 7.16 was whether we could conclude the two groups were not the same). **(b)** $\overline{x}_{EE} = 11.56$, $s_{EE} = 4.306$, $\overline{x}_{LE} = 5.12$, $s_{LE} = 4.622$. The test statistic is $t = 2.28$, df $= 8$, $P = 0.0521$. At the 0.05 level, we fail to reject the null hypothesis that the two groups have the same mean weight loss, while noting that the *P*-value is just slightly above 0.05. **(c)** With an observed difference in means of 6.44, the *P*-value for the permutation test is the proportion of times the permuted sample means have a difference at least 6.44 in absolute value. Many repetitions of the permutation test found typical *P*-values between 0.048 (where the null hypothesis would be rejected at the 0.05 level) to 0.07 (where the null would not be rejected). **(d)** The results of one set of 1000 resamples are shown below. Others will be similar. We note this BCa interval does not contain 0, so we will reject equality of the mean weight loss. Many repetitions of the boot procedures found a minimum value for the low end of the BCa interval 0.35.

```
Intervals :
Level       Normal                Basic
95%   (1.431, 11.756)      (1.596, 11.722)
Level       Percentile             BCa
95%   (1.158, 11.284)      (1.480, 11.498)
```

16.69 (a) The two populations should be the same shape, but skewed—or otherwise clearly non-Normal—so that the *t* test is not appropriate. **(b)** Either test is appropriate if the two populations are both Normal with the same standard deviation. **(c)** We can use a *t* test but not a permutation test if both populations are Normal with different standard deviations.

16.71 (a) We test $H_0: \mu = 0$ versus $H_a: \mu > 0$, where μ is the population mean difference before and after the summer language institute. We find $t = 3.86$, df $= 19$, and $P = 0.0005$. **(b)** The Normal quantile plot (right) looks odd because we have a small sample, and all differences are integers. **(c)** The *P*-value is almost always less than 0.002. Both tests lead to the same conclusion: The difference is statistically significant (the language institute did help comprehension).

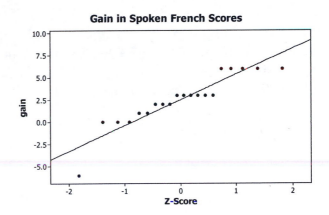

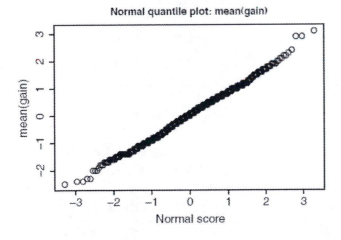

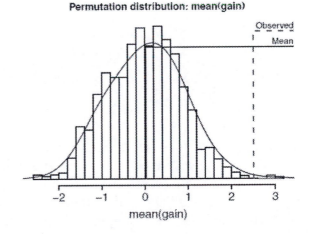

16.73 (a) We have $H_0: \rho = 0$ versus $H_a: \rho \neq 0$. **(b)** The observed correlation is $r = 0.671$. We create permutation samples and observe the proportion with correlations at least 0.671 in absolute value. You should find a *P*-value 0.002 or less. We'll conclude there is a correlation between price and rating for laundry detergents.

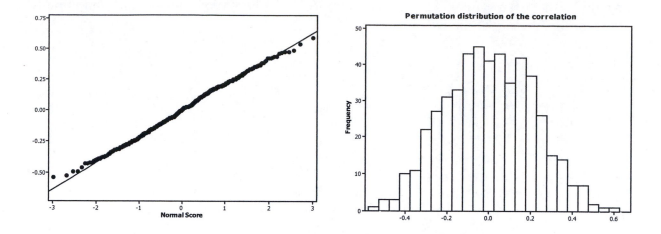

16.75 For testing $H_0 : \sigma_1 = \sigma_2$ versus $H_a : \sigma_1 \neq \sigma_2$, the permutation test P-value will almost always be between 0.065 and 0.095. In the solution to Exercise 7.107, we found $F = 1.506$ with df $= 29$ and 29, for which $P = 0.2757$—three or four times as large. In this case, the permutation test P-value is smaller, which is typical of short-tailed distributions.

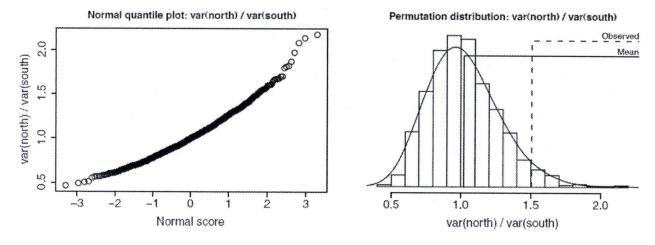

16.77 For the permutation test, we must resample in a way that is consistent with the null hypothesis. Hence, we pool the data—assuming that the two populations are the same—and draw samples (without replacement) for each group from the pooled data. For the bootstrap, we do not assume that the two populations are the same, so we sample (with replacement) from each of the two data sets separately, rather than pooling the data first.

16.79 (a) We will test $H_0 : \mu_1 = \mu_2$ versus $H_a : \mu_1 \neq \mu_2$, (males are coded as 1 in the data file). The observed mean for males was 2.7835, and the observed mean for females was 2.9325. We seek the proportion of resamples where the absolute value of the difference was at least 0.149. The P-value should generally be between 0.25 and 0.32. This test finds no significant difference in GPA between the two sexes. **(b)** We test $H_0 : \sigma_1 / \sigma_2 = 1$ versus $H_a : \sigma_1 / \sigma_2 \neq 1$. The observed ratio is $0.8593/0.7477 = 1.149$. The P-value is generally between 0.235 and 0.270. We fail to detect a difference in the standard deviations of GPAs for males and females. **(c)** Responses will vary.

16.81 (a) The standard test of $H_0 : \sigma_1 = \sigma_2$ versus $H_a : \sigma_1 \neq \sigma_2$ leads to $F = 0.3443$ with df 13 and 16; $P = 0.0587$. **(b)** The permutation P-value is typically between 0.02 and 0.03. **(c)** The P-values are similar, even though technically, the permutation test is significant at the 5% level, while the standard test is (barely) not. Because the samples are too small to assess Normality, the permutation test is safer. (In fact, the population distributions are discrete, so they cannot follow Normal distributions.)

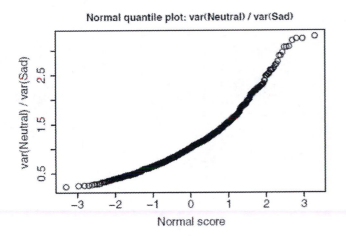

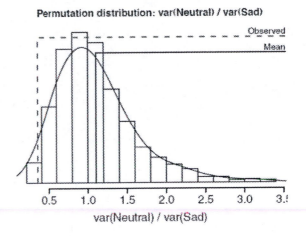

16.83 The 95% t interval is narrower in some cases than the percentile interval. Typical results are shown at right. In Example 16.8, the percentile interval was 2.793 to 3.095 (a bit narrower and higher than these) and the t interval was 2.80 to 3.10 (again, higher and narrower). This is at least in part explained by eliminating more observations on either end with the 25% trim.

Typical ranges	
t lower	2.748 to 2.757
t upper	3.012 to 3.027
Percentile lower	2.731 to 2.760
Percentile upper	3.004 to 3.028

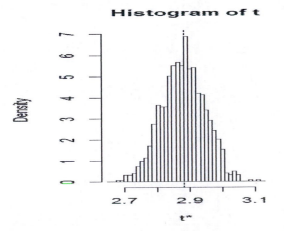

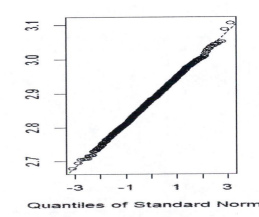

16.85 (a) The correlation for males is 0.4657. Because the bootstrap distribution does not look Normal, we'll focus on the percentile interval. The lower end of the percentile intervals ranged from 0.269 to 0.286, with the upper end ranging from 0.623 to 0.625. **(b)** The correlation for females is 0.3649. Again focusing on the percentile interval, the intervals are wider. The low end of the percentile interval ranged from 0.053 to 0.081 while the upper end ranged from 0.581 to 0.604. **(c)** Use a function like the one below to bootstrap the difference in correlations. In the plot shown, we can clearly see that the bootstrap distribution of the differences in correlations is very Normal. It is also clear that 0 will be included in the interval (the interval for this bootstrap set was $(-0.2426, 0.4229)$). All intervals examined had a low end between -0.22 and -0.24 with high end between 0.40 and 0.42. We can conclude that there is no significant difference in the correlation between high school math grades and college GPA by gender.

```
>  cd = function (D,i)  {
+ + M=subset(D[i,2:3],D[i,9]==1)
+ + F=subset(D[i,2:3],D[i,9]==2)
+ + cm=cor(M$GPA, M$HSM)
+ + cf=cor(F$GPA, F$HSM)
+ + cmf=cm-cf
+ + return (cmf)   }
```

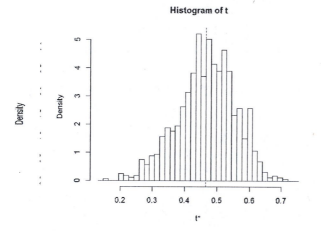

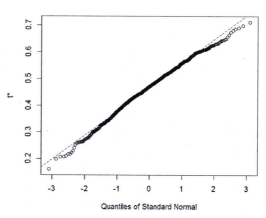

16.87 The bootstrap distribution looks quite Normal, and (as a consequence) all of the bootstrap confidence intervals are similar to each other, and also are similar to the standard (large-sample) confidence interval: 0.0981 to 0.1415.

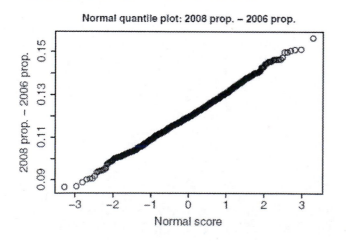

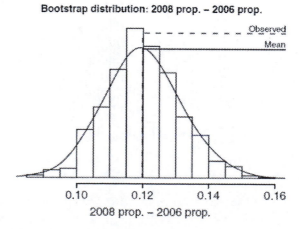

16.89 (a) There were 32 poets who died at an average age of 63.19 years ($s = 17.30$), with median age 68. There were 24 nonfiction writers, who died at an average age of 76.88 years ($s = 14.10$), with median age 77.5. In the boxplots, we can clearly see that poets seem to die younger. Both distributions are somewhat left-skewed, and the nonfiction writer who died at age 40 is a low outlier. **(b)** Using a two-sample t test, we find that testing $H_0 : \mu_N = \mu_P$ against $H_a : \mu_N \neq \mu_P$ gives $t =$ 3.26 with $P = 0.002$ (df = 53). A 95%

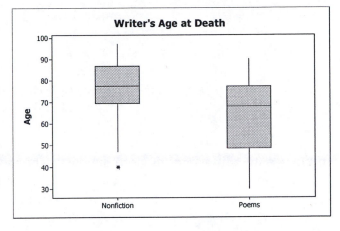

confidence interval for the difference in mean ages is (5.27, 22.11). Nonfiction writers seem to live on average between 5.22 and 22.11 years longer than poets, at 95% confidence. **(c)** The bootstrap distribution is symmetric, and seems close to Normal, except at the ends of the distribution, so a bootstrap t interval should be appropriate. The low ends of the bootstrap interval are typically between 5.26 and 5.82; the high ends are typically between 21.26 and 22.15. One particular interval seen was (5.40, 21.36). Note this interval is a bit narrower than the two-sample t interval.

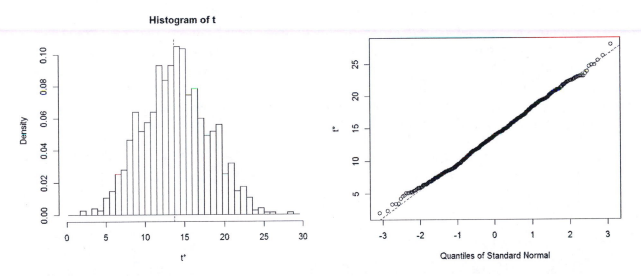

16.91 The R permutation test for the mean ages returns a P-value of 0.006 (comparable to the 0.002 from the t test) in this instance. The 99% confidence interval for the P-value is between 0.0002 and 0.0185. We can determine there is a difference in mean age at death between poets and nonfiction writers.

Exact Permutation Test Estimated by Monte Carlo

data: Age by Type1
p-value = 0.006
alternative hypothesis: true mean Type1 = 2 - mean Type1=3 is not equal to 0
sample estimates:

mean Type1=2 - mean Type1=3
-13.6875

p-value estimated from 999 Monte Carlo replications
99 percent confidence interval on p-value:
0.0002072893 0.0184986927

16.93 All answers (including the shape of the bootstrap distribution) will depend strongly on the initial sample of uniform random numbers. The median M of these initial samples will be between about 0.36 and 0.64 about 95% of the time; this is the center of the bootstrap t confidence interval. **(a)** For a uniform distribution on 0 to 1, the population median is 0.5. Most of the time, the bootstrap distribution is quite non-Normal; three examples are shown below. **(b)** SE_{boot} typically ranges from about 0.04 to 0.12 (but may vary more than that, depending on the original sample). The bootstrap t interval is therefore roughly $M \pm 2SE_{boot}$. **(c)** The more sophisticated BCa and tilting intervals may or may not be similar to the bootstrap t interval. The t interval is not appropriate because of the non-Normal shape of the bootstrap distribution and because SE_{boot} is unreliable for the sample median (it depends strongly on the sizes of the gaps between the observations near the middle).

Note: *Based on 5000 simulations of this exercise, the bootstrap t interval $M \pm 2SE_{boot}$ will capture the true median (0.5) only about 94% of the time (so it slightly underperforms its intended 95% confidence level). In the same test, both the percentile and BCa intervals included 0.5 over 95% of the time, while at the same time being narrower than the bootstrap t interval nearly two-thirds of the time. These two measures (achieved confidence level and width of confidence interval) both confirm the superiority of the other intervals. The bootstrap percentile, BCa, and tilting intervals do fairly well despite the high variability in the shape of the bootstrap distribution. They give answers similar to the exact rank-based confidence intervals obtained by inverting hypothesis tests. One variation of tilting intervals matches the exact intervals.*

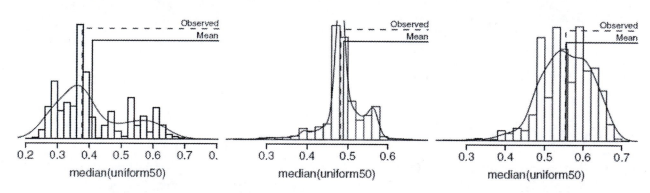

16.95 See Exercise 8.63 for more details about this survey. The bootstrap distribution appears to be close to Normal; bootstrap intervals are similar to the large-sample interval (0.3146 to 0.3854).

Typical ranges	
t lower	0.31 to 0.32
t upper	0.38 to 0.39
Percentile lower	0.30 to 0.32
Percentile upper	0.38 to 0.39

16.97 (a) This is the usual way of computing percent change: $89/54 - 1 = 1.65 - 1 = 0.65$. **(b)** Subtract 1 from the confidence interval found in Exercise 16.96; this typically gives an interval similar to 0.55 to 0.75. **(c)** Preferences will vary.

16.99 (a) The mean ratio is 1.0596; the usual t interval is $1.0596 \pm (2.262)(0.02355) = 1.0063$ to 1.1128. The bootstrap distribution for the mean is close to Normal, and the bootstrap confidence intervals (typical ranges on the right) are usually similar to the usual t interval, but slightly narrower. Bootstrapping the median produces a clearly non-Normal distribution; the bootstrap t interval should not be used for the median. (Ranges for median intervals are not given.) **(b)** The ratio of means is 1.0656; the bootstrap distribution is noticeably skewed, so the bootstrap t is not a good choice, but the other methods usually give intervals similar to 0.75 to 1.55. Also shown below is the bootstrap distribution for the ratio of the medians. It is considerably less erratic than the median ratio, but we have still not included these confidence intervals. **(c)** For example, the usual t interval from part (a) could be summarized by the statement, "On average, Jocko's estimates are 1% to 11% higher than those from other garages."

Typical ranges	
(a) Mean ratio	
t lower	1.00 to 1.02
t upper	1.10 to 1.12
Percentile lower	1.00 to 1.03
Percentile upper	1.09 to 1.11
BCa lower	1.00 to 1.03
BCa upper	1.09 to 1.11
(b) Ratio of means	
t lower	0.59 to 0.68
t upper	1.46 to 1.54
Percentile lower	0.69 to 0.78
Percentile upper	1.45 to 1.64
BCa lower	0.69 to 0.78
BCa upper	1.45 to 1.66

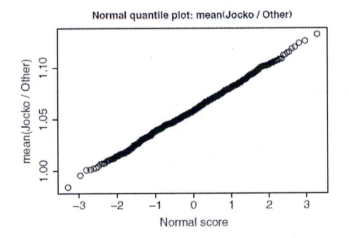

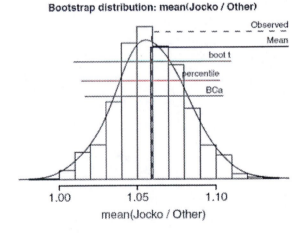

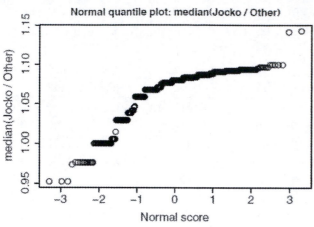

Normal quantile plot: median(Jocko / Other)

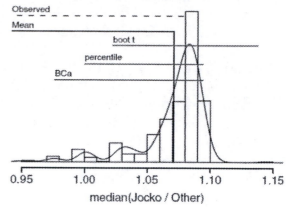

Bootstrap distribution: median(Jocko / Other)

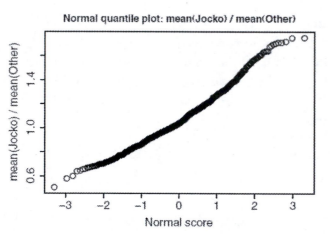

Normal quantile plot: mean(Jocko) / mean(Other)

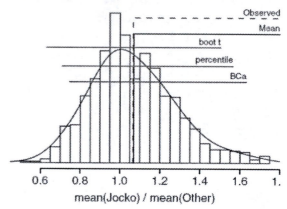

Bootstrap distribution: mean(Jocko) / mean(Other)

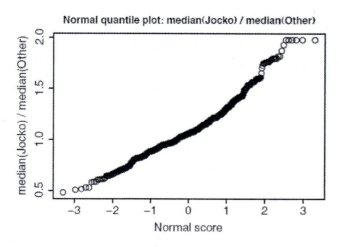

Normal quantile plot: median(Jocko) / median(Other)

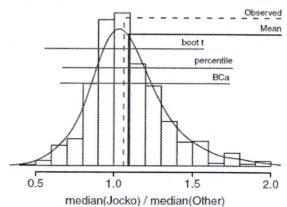

Bootstrap distribution: median(Jocko) / median(Other)

Chapter 17 Solutions

17.1 Answers will vary. For example, if brushing teeth, a possible flow chart could be

| Wet toothbrush | ➡ | Squeeze on toothpaste | ➡ | Brush teeth | ➡ | Rinse mouth | ➡ | Rinse toothbrush |

17.3 Answers will vary. Common causes for toothbrushing might be variability in the actual amount of toothpaste used each time. Special causes might include having eaten unusually sticky foods, such as caramels.

17.5 The center line is at $\mu = 90$ seconds. The control limits should be at $\mu \pm 3\sigma/\sqrt{5} = 90 \pm 3(24/\sqrt{5})$, which means about 57.8 and 122.2 seconds.

17.7 Answers here will vary but could include such steps as deciding on the order, pick-up or delivery (a yes/no), whether or not you need to look up the phone number of the shop (yes/no), actually creating the sandwich, packaging it, etc.

17.9 The most common problems are related to the application of the color coat ("orange peel," ripples in color coat, and uneven color thickness); that should be the focus of our initial efforts.

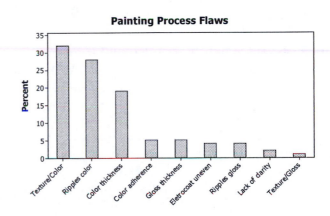

17.11 Answers will vary. Possible causes could include delivery delays due to traffic or a train, high demand during special events, how complicated the order is, and so forth.

17.13 (a) For $n = 10$, the new control limits for the $\bar{x}$ chart are $\mu \pm 3\sigma/\sqrt{10} = 1.905$ and 2.095 inches (the center line will not change). The new center line for the s chart is $0.9727(0.1) = 0.09727$ inch, and the control limits will be $B_5\sigma = (0.276)(0.1) = 0.076$ inch and $B_6\sigma = (1.669)(0.1) = 0.1669$ inches. **(b)** For $n = 2$, the new control limits for the $\bar{x}$ chart are $\mu \pm 3\sigma/\sqrt{2} = 1.7879$ and 2.212 inches. The center line for the s chart will be $(0.7979)(0.1) = 0.07979$ inch with control limits 0 and $2.606(0.1) = 0.2606$ inch. **(c)** There are 2.54 centimeters to an inch, so the centerline for the $\bar{x}$ chart is $(2)(2.54) = 5.08$ cm with control limits $(1.85)(2.54) = 4.699$ and $(2.15)(2.54) = 5.461$ cm. The new center line for the s chart will be $(0.09213)(2.54) = 0.234$ cm with control limits 0 and $(0.2088)(2.54) = 0.5304$ cm.

17.15 (a) For the $\bar{x}$ chart, the center line is at $\mu = 1.014$ lb; the control limits should be at $\mu \pm 3\sigma/\sqrt{3}$, which means about 0.9811 and 1.0469 lb. **(b)** For $n = 3$, $c_4 = 0.8862$, $B_5 = 0$, and $B_6 = 2.276$, so the center line for the s chart is $(0.8862)(0.019) = 0.01684$ lb, and the control limits are 0 and $(2.276)(0.019) = 0.04324$ lb. **(c)** The control charts are below. **(d)** The s chart is in control, but there were signals on the $\bar{x}$ chart at samples 10 and 12.

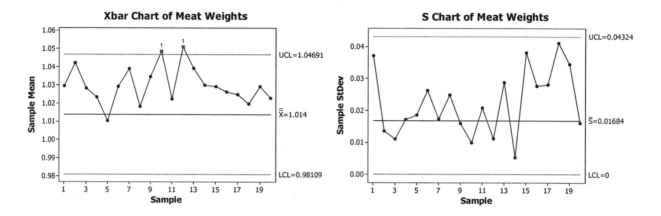

17.17 (a) The center line is at $\mu = 11.5$ Kp; the control limits should be at $\mu \pm 3\sigma\sqrt{4} = 11.5 \pm 0.3 = 11.2$ and 11.8 Kp. **(b)** Graphs below. Set A had two points out of control (the 12th sample and the last); Set C has the last two out of control. **(c)** Set B is from the in-control process. The process mean shifted suddenly for Set A; it appears to have changed on about the 11th or 12th sample. The mean drifted gradually for the process in Set C.

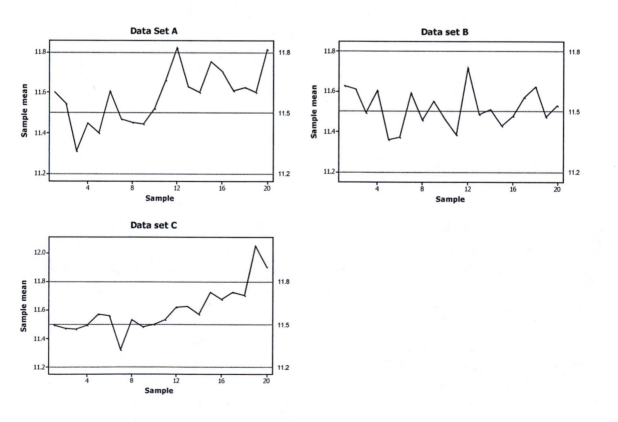

17.19 For the s chart with $n = 6$, we have $c_4 = 0.9515$, $B_5 = 0.029$ and $B_6 = 1.874$, so the center line is $(0.9515)(0.002) = 0.001903$ inch, and the control limits are $(0.029)(0.002) = 0.000058$ and $(1.874)(0.002) = 0.003748$ inch. For the $\bar{x}$ chart, the center line is $\mu = 0.85$ inch, and the control limits are $\mu \pm 3\sigma/\sqrt{6} = 0.85 \pm 0.00245 = 0.8476$ and 0.8525 inch.

17.21 For the $\bar{x}$ chart, the center line is 43, and the control limits are $\mu \pm 3\sigma/\sqrt{5} = 25.91$ and 60.09. For $n = 5$, $c_4 = 0.9400$, $B_5 = 0$, and $B_6 = 1.964$, so the center line for the s chart is $(0.9400)(12.74) = 11.9756$, and the control limits are 0 and $(1.964)(12.74) = 25.02$. The control charts (below) show that sample 5 was above the UCL on the s chart, but it appears to have been a special cause variation, as there is no indication that the samples that followed it were out of control.

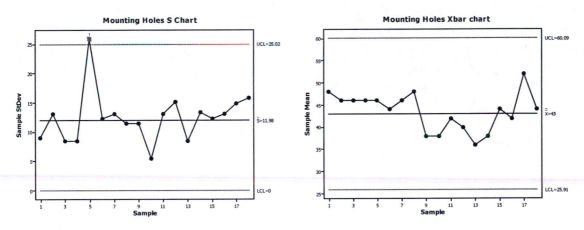

17.23 (a) The process mean is the same as the center line: $\mu = 715$. The control limits are three standard errors from the mean, so $35 = 3\sigma/\sqrt{4}$, meaning that $\sigma = 23.3333$. **(b)** If the mean changes to $\mu = 700$, then $\bar{x}$ is approximately Normal with mean 700 and standard deviation $\sigma/\sqrt{4} = 11.6667$, so $\bar{x}$ will fall outside the control limits with probability $1 - P(680 < \bar{x} < 750) = 1 - P(-1.71 < Z < 4.29) = 0.0436$. **(c)** With $\mu = 700$ and $\sigma = 10$, $\bar{x}$ is approximately Normal with mean 700 and standard deviation $\sigma/\sqrt{4} = 5$, so $\bar{x}$ will fall outside the control limits with probability $1 - P(680 < x < 750) = 1 - P(-4 < Z < 10) = 1$.

17.25 The usual 3σ limits are $\mu \pm 3\sigma/\sqrt{n}$ for an $\bar{x}$ chart and $(c_4 \pm 3c_5)\sigma$ for an s chart. For 2σ limits, simply replace "3" with "2." **(a)** $\mu \pm 2\sigma/\sqrt{n}$. **(b)** $(c_4 \pm 2c_5)\sigma$.

17.27 (a) Shrinking the control limits would increase the frequency of false alarms because the probability of an out-of-control point when the process is in control will be higher (roughly 5% instead of 0.3%). **(b)** Quicker response comes at the cost of more false alarms. **(c)** The runs rule is better at detecting gradual changes. (The one-point-out rule is generally better for sudden, large changes.)

17.29 We estimate $\hat{\sigma}$ to be $s/c_4 = 1.03/0.9213 = 1.1180$, so the $\bar{x}$ chart has center line $\bar{\bar{x}} = 47.2$ and control limits $\bar{x} \pm 3\,\hat{\sigma}/\sqrt{4} = 45.523$ and 48.877. The s chart has center line $\bar{s} = 1.03$ and control limits 0 and $(2.088)(\hat{\sigma}) = 2.3344$.

17.31 (a) The centerline for the s chart will be at $\bar{s} = 0.07177$. In Exercise 17.12, we had $n = 4$ for each sample, so $c_4 = 0.9213$ and $\hat{\sigma} = \bar{s}/c_4 = 0.07177/0.9213 = 0.07790$. We'll have UCL $=$ $2.088(0.07790) = 0.16266$ and LCL $= 0$. **(b)** None of the sample standard deviations were greater than 0.16266, so variability is under control. **(c)** The centerline for the $\bar{x}$ chart is $\bar{\bar{x}} = 2.0078$. The limits are $\bar{x} \pm 3\,\hat{\sigma}/\sqrt{4} = 2.0078 \pm 0.10766 = 1.90014$ to 2.11546 inches. **(d)** These limits are slightly narrower than those found in Exercise 17.12 (1.85 to 2.15 inches).

17.33 Sketches will vary. One possible x chart is shown, created with the (arbitrary) assumption that the experienced clerk processes invoices in an average of 2 minutes, while the new hire takes an average of 4 minutes. (The control limits were set arbitrarily as well.)

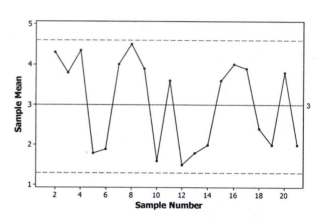

    Note: *Such a process would not be considered to be in control for very long. The initial control limits might be developed based on a historical estimate of σ, but eventually we should assess that estimate based on our sample standard deviations. Because both clerks "are quite consistent, so that their times vary little from invoice to invoice," each sample has a small value of s, so the revised estimate of σ would likely be smaller. At that point, the control limits (based on that smaller spread) will be moved closer to the center line.*

17.35 (a) Average the 20 sample means and standard deviations, and estimate μ to be $\hat{\mu} = \bar{\bar{x}} = 2750.7$ and σ to be $\hat{\sigma} = \bar{s}/c_4 = 345.5/0.9213 = 375.0$. **(b)** In the s chart shown in Figure 17.7, most of the points fall below the center line.

17.37 If the manufacturer practices SPC, that provides some assurance that the phones are roughly uniform in quality—as the text says, "We know what to expect in the finished product." So, assuming that uniform quality is sufficiently high, the purchaser does not need to inspect the phones as they arrive because SPC has already achieved the goal of that inspection, to avoid buying many faulty phones. (Of course, a few unacceptable phones may be produced and sold even when SPC is practiced—but inspection would not catch all such phones anyway.)

17.39 The quantile plot does not suggest any serious deviations from Normality, so the natural tolerances should be reasonably trustworthy.

    Note: *We might also assess Normality with a histogram or stemplot; this looks*

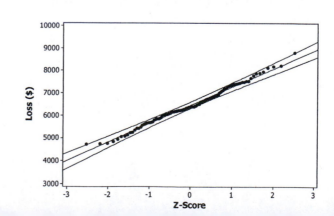

reasonably Normal, but we see that the number of losses between $6000 and $6500 is noticeably higher than we might expect from a Normal distribution. In fact, the smallest and largest losses were $4727 and $8794. These are both within the tolerances, but note that the minimum is quite a bit more than the lower limit of the tolerances ($4008). The large number of losses between $6000 and $6500 makes the mean slightly lower and therefore lowers both of the tolerance limits.

17.41 If we shift the process mean to 2500 mm, about 99% will meet the new specifications:

$$P(1500 < X < 3500) = P\left(\frac{1500 - 2500}{384} < Z < \frac{3500 - 2500}{384}\right) = P(-2.60 < Z < 2.60) = 0.9953 -$$

$0.0047 = 0.9906.$

17.43 The mean of the 17 in-control samples is $\bar{\bar{x}} = 43.4118$, and the standard deviation is 11.5833, so the natural tolerances are $\bar{\bar{x}} \pm 3s = 8.66$ to 78.16.

17.45 Only about 44% of meters meet the specifications. Using the mean (43.4118) and standard deviation (11.5833) found in the solution to Exercise 17.43:

$$P(44 < X < 64) = P\left(\frac{44 - 43.4118}{11.5833} < Z < \frac{64 - 43.4118}{11.5833}\right) = P(0.05 < Z < 1.78) = 0.9625 - 0.5199 =$$

0.4426

17.47 The limited precision of the measurements shows up in the granularity (stair-step appearance) of the graph (right). Aside from this, there is no particular departure from Normality.

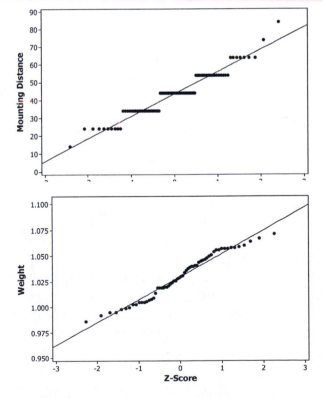

17.49 The quantile plot, while not perfectly linear, does not suggest any serious deviations from Normality, so the natural tolerances should be reasonably trustworthy.

17.51 (a) (ii) A sudden change in the $\bar{x}$ chart: This would immediately increase the amount of time required to complete the checks. **(b)** (i) A sudden change (decrease) in s or R because the new measurement system will remove (or decrease) the variability introduced by human error.

(c) (iii) A gradual drift in the $\bar{x}$ chart (presumably a drift up, if the variable being tracked is the length of time to complete a set of invoices).

17.53 The process is no longer the same as it was during the downward trend (from the 1950s into the 1980s). In particular, including those years in the data used to establish the control limits results in a mean that is too high to use for current winning times, and a standard deviation that includes variation attributable to the "special cause" of the changing conditioning and professional status of the best runners. Such special cause variation should not be included in a control chart.

17.55 LSL and USL are specification limits on the individual observations. This means that they do not apply to averages and that they are *specified* as desired output levels, rather than being *computed* based on observation of the process. LCL and UCL are control limits for the averages of samples drawn from the process. They may be determined from past data, or independently specified, but the main distinction is that the purpose of control limits is to detect whether the process is functioning "as usual," while specification limits are used to determine what percentage of the outputs meet certain specifications (are acceptable for use).

17.57 For computing $\hat{C}_{pk}$, note that the estimated process mean (2750.7 mm) lies closer to the USL. **(a)** $\hat{C}_p = \dfrac{4000 - 1000}{6 \times 383.8} = 1.3028$ and $\hat{C}_{pk} = \dfrac{4000 - 2750.7}{3 \times 383.8} = 1.0850$. **(b)** $\hat{C}_p = \dfrac{3500 - 1500}{6 \times 383.8} = 0.8685$ and $\hat{C}_{pk} = \dfrac{3500 - 2750.7}{3 \times 383.8} = 0.6508$.

17.59 In the solution to Exercise 17.48, we found that the mean and standard deviation of all 60 weights are $\bar{x} = 1.02997$ lb and $s = 0.0224$ lb. **(a)** $\hat{C}_p = \dfrac{1.10 - 0.96}{6 \times 0.0224} = 1.0417$ and $\hat{C}_{pk} = \dfrac{1.10 - 1.03}{3 \times 0.0224} = 1.0418$. (These were computed with the unrounded values of $\bar{x}$ and s; rounding will produce slightly different results.) **(b)** Customers typically will not complain about a package that was too heavy.

17.61 (a) $C_{pk} = \dfrac{0.75 - 0.25}{3\sigma} = 0.5767$. 50% of the output meets the specifications. **(b)** LSL and USL are 0.865 standard deviations above and below the mean, so the proportion meeting specifications is $P(-0.865 < Z < 0.865) = 0.6130$. **(c)** The relationship between C_{pk} and the proportion of the output meeting specifications depends on the shape of the distribution.

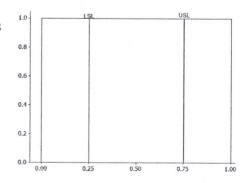

17.63 See also the solution to Exercise 17.47. **(a)** Use the mean and standard deviation of the 85 remaining observations: $\hat{\mu} = \bar{x} = 43.4118$ and $\hat{\sigma} = s = 11.5833$.

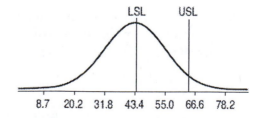

(b) $\hat{C}_p = \dfrac{64 - 44}{6\hat{\sigma}} = 0.2878$ and $\hat{C}_{pk} = 0$ (because $\hat{\mu}$ is outside the specification limits). This process has very poor capability: The mean is too low and the spread too great. Only about 46% of the process output meets specifications.

17.65 We have $\bar{x} = 7.996$ mm and $s = 0.0023$ mm, so we assume that an individual bearing diameter X follows a $N(7.996, 0.0023)$ distribution. **(a)** About 91.8% meet specifications:

$$P(7.992 < X < 8.000) = P\left(\frac{7.992 - 7.996}{0.0023} < Z < \frac{8.000 - 7.996}{0.0023}\right) = P(-1.74 < Z < 1.74) = 0.9591 -$$

$0.0409 = 0.9182.$ **(b)** $\hat{C}_{pk} = \dfrac{8.000 - 7.996}{3 \times 0.0023} = 0.5797.$

17.67 This graph shows a process with Normal output and $C_p = 2$. The tick marks are σ units apart; this is called "six-sigma quality" because the specification limits are (at least) six standard deviations above and below the mean.

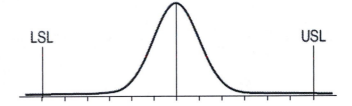

17.69 Students will have varying justifications for the sampling choice. Choosing six calls per shift gives an idea of the variability and mean for the shift as a whole. If we took six consecutive calls (at a randomly chosen time), we might see additional variability in $\bar{x}$ because sometimes those six calls might be observed at particularly busy times (when a customer has to wait for a long time until a representative is available or when a representative is using the restroom).

17.71 The outliers are 276 seconds (sample 28), 244 seconds (sample 42), and 333 seconds (sample 46). After dropping those outliers, the standard deviations drop to 9.284, 6.708, and 31.011 seconds. (Sample #39, the other out-of-control point, has two moderately large times, 144 and 109 seconds; if they are removed, s drops to 3.416.)

17.73 (a) For those 10 months, there were 957 overdue invoices out of 26,350 total invoices (opportunities), so $\bar{p} = \dfrac{957}{26,350} = 0.03632.$ **(b)** The center line and control limits are: $\mathrm{CL} = \bar{p} = $

$0.03632,$ control limits: $\bar{p} \pm 3\sqrt{\dfrac{\bar{p}(1 - \bar{p})}{2635}} = 0.02539$ and $0.04725.$

17.75 The center line is at the historical rate (0.0231); the control limits are

$0.0231 \pm 3\sqrt{\dfrac{0.0231(0.9769)}{500}}$, which means about 0.00295 and 0.04325.

17.77 The center line is at $\bar{p} = \dfrac{194}{38,370} = 0.00506$; the control limits should be at $\bar{p} \pm 3\sqrt{\dfrac{\bar{p}(1 - \bar{p})}{1520}}$,

which means about -0.0004 (use 0) and 0.01052.

17.79 (a) The student counts sum to 9218, while the absentee total is 3277, so $\bar{p} = \dfrac{3277}{9218}$ = 0.3555 and n = 921.8. **(b)** The center line is $\bar{p}$ = 0.3555, and the control limits are:

$$\bar{p} \pm 3\sqrt{\frac{\bar{p}(1-\bar{p})}{921.8}} = 0.3082 \text{ and } 0.4028$$

The p chart suggests that absentee rates are in control. **(c)** For October, the limits are 0.3088 and 0.4022; for June, they are 0.3072 and 0.4038. These limits appear as solid lines on the p chart, but they are not substantially different from the control limits found in (b). Unless n varies *a lot* from sample to sample, it is sufficient to use n.

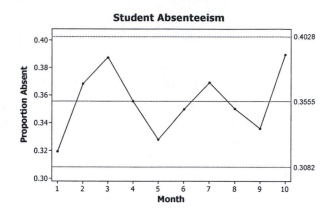

17.81 (a) $\bar{p} = \dfrac{8,000}{1,000,000}$ = 0.008. We expect about (500)(0.008) = 4 defective orders per month.

(b) The center line and control limits are: CL = $\bar{p}$ = 0.008, control limits: $\bar{p} \pm 3\sqrt{\dfrac{\bar{p}(1-\bar{p})}{500}}$ = −0.00395 and 0.01995 (We take the lower control limit to be 0.) It takes at least ten bad orders in a month to be out of control because (500)(0.01995) = 9.975.

17.83 (a) The percents do not add to 100% because one customer might have several complaints; that is, he or she could be counted in several categories. **(b)** Clearly, top priority should be given to the processes of ease of adjustment/credits, as well as both the accuracy and the completion of invoices, as the three most common complaints involved invoices.

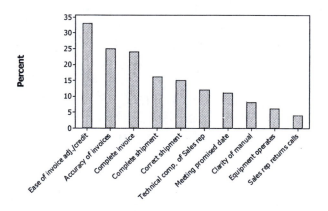

17.85 On one level, these two events are similar: Points above the UCL on an $\bar{x}$ (s) chart suggest that the process mean (standard deviation) may have increased. The difference is in the implications of such an increase (if not due to a special cause). For the mean, an increase might signal a need to recalibrate the process in order to keep meeting specifications (that is, to bring the process back into control). An increase in the standard deviation, on the other hand, typically does not indicate that adjustment or recalibration is necessary, but it will require re-computation of the $\bar{x}$ chart control limits.

17.87 We find that $\bar{s} = 7.65$, so with $c_4 = 0.8862$ and $B_6 = 2.276$, we compute $\hat{\sigma} = \bar{s}/c_4 = 7.65/0.8862 = 8.63$, LCL = 0, and UCL = (2.276)(8.63) = 19.65. One point (from sample #1) is out of control. (And, if that cause were determined and the point removed, a new chart would have s for sample #10 out of control.) The second (lower) UCL line on the control chart is the final UCL, after removing both of those samples (per the instructions in Exercise 17.88).

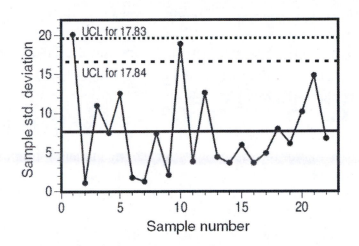

17.89 (a) As was found in the previous exercise, $\hat{\sigma} = \bar{s}/c_4 = 7.295$. Therefore, $C_p = \dfrac{50}{6\hat{\sigma}} = 1.1423$. This is a fairly small value of C_p; the specification limits are just barely wider than the $6\hat{\sigma}$ width of the process distribution, so if the mean wanders too far from 830, the capability will drop. **(b)** If we adjust the mean to be close to 830 mm $\times 10^{-4}$ (the center of the specification limits), we will maximize C_{pk}. C_{pk} is more useful when the mean is not in the center of the specification limits. **(c)** The value of $\hat{\sigma}$ used for determining C_p was estimated from the values of s from our control samples. These are for estimating short-term variation (within those samples) rather than the overall process variation. To get a better estimate of the latter, we should instead compute the standard deviation s of the *individual* measurements used to obtain the means and standard deviations given in Table 17.14 (specifically, the 60 measurements remaining after dropping samples 1 and 10). These numbers are not available. (See "How to cheat on C_{pk}" on page 17–46 of Chapter 17.)

17.91 (a) Use a p chart, with center line $\bar{p} = \dfrac{15}{5000} = 0.003$ and control $\bar{p} \pm 3\sqrt{\dfrac{\bar{p}(1-\bar{p})}{100}}$, or 0 to 0.0194. **(b)** There is little useful information to be gained from keeping a p chart: If the proportion remains at 0.003, about 74% of samples will yield a proportion of 0, and about 22% of proportions will be 0.01. To call the process out of control, we would need to see two or more unsatisfactory films in a sample of 100.

17.93 Several interpretations of this problem are possible, but for most reasonable interpretations, the probability is about 0.3%. From the description, it seems reasonable to assume that all three points are inside the control limits; otherwise, the one-point-out rule would take effect. Furthermore, the phrase "two out of three" could be taken to mean either "*exactly* two out of three," or "*at least* two out of three." (Given what we are trying to detect, the latter makes more sense, but students may have other ideas.) For the kth point, we name the following events:

- A_k = "that point is no more than $2\sigma/\sqrt{n}$ from the center line,"
- B_k = "that point is 2 to 3 standard errors from the center line."

For an in-control process, $P(A_k) = 95\%$ (or 95.45%) and $P(B_k) = 4.7\%$ (or 4.28%). The first given probability is based on the 68–95–99.7 rule; the second probability (in parentheses) comes from Table A or software. Note that, for example, the probability that the first point gives no cause for concern, but the second and third are more than $2\sigma / \sqrt{n}$ from, and on the same side of, the center line, would be:

$$\frac{1}{2} P(A_1 \cap B_2 \cap B_3) = 0.10\% \text{ (or 0.09\%)}$$

(The factor of 1/2 accounts for the second and third points being on the same side of the center line.) If the "other" point is the second or third point, this probability is the same, so if we interpret "two out of three" as meaning "*exactly* two out of three," then the total probability is three times the above number:

$P(\text{false out-of-control signal from an in-control process}) = 0.31\% \text{ (or 0.26\%)}$

With the (more-reasonable) interpretation "*at least* two out of three":

$P(\text{false out-of-control signal}) = \frac{1}{2} P(A_1 \cap B_2 \cap B_3) + \frac{1}{2} P(B_1 \cap A_2 \cap B_3) + \frac{1}{2} P(B_1 \cap B_2) =$

$0.32\% \text{ (or 0.27\%)}$